D1357239

IDENTICALLY DIFFERENT

Tim Spector

IDENTICALLY DIFFERENT

Why you *can* change your genes

WEIDENFELD & NICOLSON
LONDON

First published in Great Britain in 2012
by Weidenfeld & Nicolson

1 3 5 7 9 10 8 6 4 2

A CIP catalogue record for this book is
available from the British Library.

ISBN: 978 0 297 86631 2

Typeset by Input Data Services Ltd, Bridgwater, Somerset

Printed and bound by CPI Group (UK) Ltd, Croydon, CR0 4YY

The Orion Publishing Group's policy is to use papers that are natural,
renewable and recyclable and made from wood grown in sustainable
forests. The logging and manufacturing processes are expected to
conform to environmental regulations of the country of origin.

Weidenfeld & Nicolson

Orion Publishing Group Ltd
Orion House
5 Upper Saint Martin's Lane
London, WC2H 9EA

A Hachette UK Company

www.orionbooks.co.uk

Some of the names, dates and places related to the
personal twin stories in the book have been altered
to protect the privacy of the individuals.

For Veronique, Henri and Phillip

CONTENTS

Preface 1

Introduction: Was Darwin wrong? 7

1 The gene myth 27
Toads, giraffes and fraud

2 'The happiness gene' 45
Mindsets, optimism, laughter

3 'The talent gene' 66
Genius, motivation and taxi drivers

4 'The God gene' 88
Genesis, Jedi and Hollywood

5 'The parenting gene' 110
Nature, nurture and naughtiness

6 'Bad genes' 130
Abusers, criminals and victims

7 'The mortality gene' 153
Hearts, famines and grandparents

8 'The fat gene' 166
Diet, worms and wine

9	**'The cancer gene'** Autism, toxins and babies	192
10	**'The gay gene'** Sex, hormones and the brain	220
11	**'The fidelity gene'** Pleasure, pain and the G spot	240
12	**Bacteria genes** Bugs, poo and you	260
13	**Identical genes** Clones, identity and the future	278
	Acknowledgements	294
	Notes	296
	Index	316

PREFACE

The procedure had been planned painstakingly for months, yet the operation, as it reached the crucial stage, was not going well.[1] After many hours on their feet performing intricate vascular brain surgery, the team was tiring fast. A small severed artery pumped a thin stream of blood onto one of the surgeon's glasses before it was sealed off. Some other blood vessels were leaking steadily, and it was not clear what caused the bleeding and where exactly it originated. Clamping a number of suspect veins and arteries didn't work, the area of the back of the exposed brain just kept filling up with fresh blood and obscuring everything, making it hard to distinguish the grey shiny brain tissue from the vessels.

'The blood pressure's dropping,' said the chief anaesthetist.

'Not good', thought Dr Keith Goh, the head surgeon, grimly. Then: 'Let's give the rest of the blood transfusion now and get some more units up here. We'll try more compression and stitches and reseal the dura and see if it holds. I don't think we have any other choice.'

Twenty-nine years earlier and several thousand miles away, two baby girls, identical twins, had been conceived as the result of a series of chance events. At just the right day of the month, one of their father's many billion sperm containing a half-set of 23 chromosomes met and fertilised one of their mother's 400 eggs containing another half-set of 23 chromosomes. A few days later a single fertilised egg, still containing no more than a handful of cells, suddenly split, and produced two genetically identical embryos. The two baby clones developed for nine months side by side.

The twins were born on a cold day in January in the troubled years just before the revolution in Firouzabad in southwest Iran. At first they hardly saw their parents, who were poor farmers with nine other kids to feed and look after. Because of complications the twins stayed in hospital, and because of their parents' financial problems they were adopted by a kindly doctor.

The two girls did everything together, eating, playing, sleeping, and never left each other's side. Despite their identical genes and environment, there were obvious differences between them. Ladan liked animals, whereas Laleh preferred computer games, which Ladan, who preferred to pray, couldn't stand. When they were older, both started to enjoy shopping, particularly for cosmetics. Ladan was left-handed and Laleh right-handed. They did well at school, though they often whispered answers to each other in exams. They wanted to continue to study together, but Ladan hoped to be a lawyer in Tehran and Laleh a journalist in Shiraz. Eventually Ladan won the argument and they both studied law in Tehran. When asked, they would both agree that Ladan was the talkative extrovert and Laleh was more introverted.

How could the differences in personality of these two girls be explained? They were genetic clones with exactly the same DNA structure and every one of the 100 trillion cells of their body contained the same 25,000 genes.[2] They had been attended to, and initially fed, by the same mother, and later brought up by the same adopted father. They had spent every day of their lives together, gone to the same school and university; they had the same friends and the same diet. They also had a special and unique bond: they were literally inseparable. They were Siamese twins conjoined at the head.

As the twins got older, their desire for independence grew, and they spent six years trying to convince doctors to perform surgery to separate them. All the experienced doctors they consulted declined, because of the very high probability of death in such a complex operation. The twins shared the major vein (the sagittal sinus) running behind the brain which acts as the main reservoir of blood. In 2003 they finally convinced Dr Goh, a senior neurosurgeon from Singapore, to operate on them despite the apparent risks. He had performed previous successful operations on younger twins and optimistically put the risk of death at closer to fifty-fifty.

The operation started on a humid morning in July in Singapore with a confirmatory MRI scan. It lasted 52 hours and involved 28 surgeons from four countries, as well as hundreds of support staff. The operation itself cost millions of dollars – they had to use a specially designed operating table resembling a double dental chair – and Iranian TV crews gave regular updates.

After hesitating at a crucial stage when the operation was nearly aborted, the team finally managed to separate the heads. But the brains were more closely fused than they expected from the scans; bleeding from the shared complex network of blood vessels could not be controlled. Neither of the twins regained consciousness, and despite the best efforts of the team, they died soon after. A note written by them and posted on the hospital website the day of the operation read: 'We have been praying every day for our operation . . . We hope the operation will finally bring us to the end of this difficult path, and we may begin our new and wonderful lives as two separate persons.' The twins got their wish. They were finally separated in death – they were buried in individual tombs – yet still lie side by side.

Conjoined twins are thankfully extremely rare (1 in 2 million births), but the story of Ladan and Laleh illustrates a key point. Most of us share very similar genes and environments to our siblings and parents, yet our personalities, tastes, physical appearance and health turn out to be entirely different. If our genes and environment are the same, how can there be any room for differences between us? And, if so, how do these differences arise?

In 2009 I undertook the role of scientific consultant for a BBC two-part series called *The Secret Life of Twins*. The first programme was straightforward to plan, as it presented case studies on how uncannily similar some identical twins are, even those that have been separated at birth. There was a great example of two Chinese girls, Mia and Alexandra, separated as infants and adopted by two families in Sacramento, California and the fjords of Norway respectively. Each family was unaware of the other twin's existence.

On the occasion of the filming of our documentary, Mia and Alexandra, aged six, were reunited. Although they lacked a common language they became instant friends. From the time they first skipped out of the car and waved to each other it was clear they had seemingly identical habits and mannerisms. Their brief time together, and the traumatic reseparation that followed, didn't leave a dry eye in the house.

I explained to the producers of the programme that, although fascinating to the public, the example of these twins was less stimulating for scientists, as it had been well documented in the past that identical twins raised apart develop many parallels and striking similarities. The producers challenged me to come up with something more novel and provocative for the second programme. I suggested that they look at the complete opposite scenario to Mia and Alexandra: namely, identical twins raised together who end up very different. Scientists have plausible explanations for identical twins like Mia and Alexandra turning out very much alike, as we now know that most traits and characteristics are at least partly influenced by our genes. But we have no idea why people who share the same genes and similar environments can and do turn out to be very different (or discordant, to use the scientific term).

The case histories we eventually chose ranged from autism, obesity, anorexia and skin ageing to homosexuality. Although discussed only briefly, these few cases had a major impact. They shattered our comfortable perceptions of determinism and individuality. Rather than looking at these pairs as rare oddities, I realised that these human stories had changed my views and offered the key to unlocking a wider understanding of ourselves and reaching an alternative view of genetic 'determinism'.

The example of Ladan and Laleh – with identical genes and environments yet very different personalities – shows us the limits of our understanding. This book explores this new way of looking at genes. It forces us to reconsider key questions of who we are and why, and how we explain our behaviour, characteristics and diseases. Armed with this new science [3] we may even have to rewrite some of the Darwinian principles that have dominated scientific thinking for so long. With these new ideas we can begin to understand the fundamental question of what makes us all so alike and yet so different.

INTRODUCTION

Was Darwin wrong?

Until three years ago I was one of the many scientists who took the gene-centric view of the universe for granted. I had spent the last 17 years producing hundreds of twin studies trying to convince a sceptical public and scientific world that virtually every trait and disease had a major genetic influence. My colleagues and I around the world were largely successful in this, and the prospect of finding the genes underlying most diseases looked increasingly certain. But I had a nagging doubt that we were missing something.

Scientific dogma has long stated that genes are fixed entities and cannot be changed. Once we have inherited them, they can't be affected by our environment and remain with us until we die or pass them on – unchanged – to the next generation. While we can influence our lives by choosing our friends, spouses, lifestyles, or training our memory, our genes are always immutable. Genes are viewed as central to how the body and cells develop and work, our 'blueprint', 'the code of life' – or so it was thought.

By studying twins we found that for just about every disease we looked at, identical twins (who share the same genes) both developed the disease more often than non-identical (fraternal) twins (who share only half their genes). The degree of sharing is called a correlation, and by a bit of simple maths comparing these correlations we can produce a measure of the genetic fraction called heritability.

For example, if you were measuring the heritability of weight you might compare the weights of 50 identical twin pairs with 50 non-identical twin pairs by adding up the similarities in one group with the similarities in the other. If the average similarity in identical twins is 90 per cent and in non-identicals is 60 per cent, the heritability is found by doubling the difference between them, i.e. 30 × 2. So we would say that weight in this example is 60 per cent heritable. Calculating the heritability of diseases is slightly more complicated but the principle is the same. This is simply the proportion of differences between people explained by genes.

Diseases like rheumatoid arthritis have heritabilities of 60–70 per cent, so appear to be strongly genetic. Yet when we looked at identical women twins with the disease, 85 per cent of the women never developed their sister's disease – even though they had the same genes and very similar lifestyles.[1] I found this same pattern was true for most diseases studied: there was rarely more than a 50 per cent chance of both twins getting the same disease, and usually the figure was much lower. This was a worry: I realised that my traditional view of genetics and the dominant role of genes might have to change.

Just over ten years ago researchers found that the diets of pregnant mothers could alter the behaviour of genes in their children and that these changes could last a lifetime and then be passed on in turn to their children. The genes were literally being switched on or off by a new mechanism we call epigenetics – meaning in Greek 'around the gene'. Contrary to traditional genetic dogma, these changes could be transferred to the next generation. In this case the mothers just happened to be rats, but recent similar findings in humans have created a revolution in our thinking.

Darwin's theory of natural selection and evolution, published

150 years ago, was based on a number of simple but broad concepts that have since been frequently refined and sometimes misquoted.[2] His theory was the result of a more general diffuse set of ideas than commonly appreciated, based on the laws of reproduction, inheritance, variability between individuals, and a struggle for survival. The slow process of natural selection will occur in a world in which organisms can reproduce themselves, and there are differences (variation) between individuals. When these individuals reproduce they pass on characteristics of the parents, and these inherited characteristics affect the offspring's success in survival and reproduction. A key factor in the process of natural selection is that it is blind and driven by random variations. Darwin himself knew nothing of genes, Mendel's laws of inheritance or DNA, all of which would only become attached to the theory of evolution in the next century.

The gene-centric view is in fact a relatively new phenomenon, firmed up by the coming together of a number of discoveries in the early twentieth century.[3] These included finding that genes are segments of DNA that code for proteins, the chemicals that drive all the body's reactions, made up of a number of amino acids put together in the cell. They also found that these genes come in pairs, each called alleles, which are lined up along 23 paired chromosomes (strands like pipe-cleaners) in each cell in the body. One pair of alleles is inherited from the father and one from the mother. These pairs conveniently split apart when sperm or egg cells are made, so that each contains only one half of the 46 chromosomes, and therefore when they fuse to make a fetus the number of chromosomes and genes stays the same. The splitting and fusing also involves randomly shuffling the (unchanged) parents' genes, so that no two eggs or sperm contain the same combinations of genes.

James Watson and Francis Crick in 1953 worked out that

DNA was a double helix made up of four interlocking chemical bases (abbreviated as T,A,G,C), which zips and unzips – a process that explains why the copies of DNA and genes are so reliable. It was found that these bases lined up opposite each other in a complementary way, which was always the same. This led to the discovery of a smaller molecule called RNA, which translates the DNA code to make the proteins (and so the enzymes) that drive the cell. It was also discovered that genes could, in rare cases, spontaneously 'mutate', causing diseases and traits like dwarfism, and this random event was believed to be one source of the natural variation that could be inherited.

The gene was the key. All these insights and the molecular biology and discoveries that followed tended to focus on the pivotal role of the gene as being the primary driver, causing its effects via the proteins. No one asked whether the environment would be influencing genes or proteins, or whether proteins might influence genes in the opposite direction. The fact that Darwin included a role for acquired inheritance in his theory of evolution is often overlooked.

However, Darwin's big idea was that the main force of evolution was by random selection of the fittest elements of each generation, working over millennia. This (so-called 'survival of the fittest') theory perfectly explains our genetic similarity to other species. Whereas all humans share all of their approximately 25,000 genes (the current estimated number is close to 23,000, but this number is constantly being readjusted) with each other, they do vary in which variants of each of the genes are shared, so that whereas siblings share the same basic genes, they will have on average only 50 per cent of the variable bits in common. In all first-degree relatives (siblings and parents) some gene variants will be shared identically, others not at all. Our closeness in evolutionary time to our shared ancestors is exactly mirrored by our similarities in DNA. So we humans share 99 per cent of the same DNA sequence as chimpanzees, from whom we split 6 million years ago, 90 per cent with mice (100 million years), and even 31 per cent with yeast (1.5 billion years). This close genetic relationship with chimpanzees is an uncomfortable fact that creationists have great trouble explaining away, other than by invoking God's playful nature – an effort to confuse us.

In 2000 Bill Clinton and Tony Blair proudly unveiled to the world one of the great advances for mankind: the sequencing of the 3 billion base pairs that make up the DNA in every cell of a single human. Within all this DNA, the sequence of all the genes was suddenly laid bare. Now, we thought, we had the tools to unlock how humans and animals worked. At the time it was billed as 'the opening of the book of life'. All sorts of scientific and medical breakthroughs were predicted to follow.

The twentieth century may have been the century of the gene, but genetics has made astonishing advances in the twenty-first century. While the sequencing of the 3 billion bases of genetic code of the first human cost over $2,000,000,000 dollars and involved thousands of scientists working for over ten years, it can now be done for $2,000 and falling – a million-fold discount in ten years. This revolutionary technology has had many other spinoffs.

The Spector genes

Curious to discover what the new technology could tell me about my own genetic health risks and family roots, I wanted to learn at first hand what the Internet could offer without seeing a specialist. I had heard of a couple of companies promising personalised genomics tests – Decode, based in Iceland, and 23andMe, in the US. Being a sceptical type of person I applied to both companies independently so that I could compare them. After paying a few hundred dollars for an ancestry and health check I received in the post a tube to spit into for several minutes from one company, and from the other a wooden stick to rub on the inside of my cheek.[4]

The companies extracted my DNA from the cells of my saliva or cheek tissue and then measured nearly 1 million genetic markers (called SNPs and pronounced 'snips') on each sample. They would then match them up with reports from published studies (some of which I had co-authored) that link these markers with diseases, personal traits and your ancestral origins.

Two weeks later the results arrived back via the Internet. They came with plenty of dire warnings about the consequences of knowing the results. Could they change my life? I had heard of journalists trembling when they opened the results, fearing the worst. What about me? Would I be doomed to cancer or Alzheimer's? After some mild apprehension, in the end it wasn't so bad, and the results were generally well explained with lots of warnings.

Consulting the findings from 23andMe, I decided to look at my ancestry first. This showed that I was 30 per cent North European, 60 per cent South European and unexpectedly 10 per cent Asian. This was reassuring, as results were similar between the companies. I get a good suntan, but I still can't explain where the recent Asian genes came from, although I knew it was from my mother's side of the family, who is supposedly white Australian.

I then checked the disease results and was relieved to see that out of the 20 or so diseases listed, my risk for most of them was low. I did however have a worrying increased risk for diabetes, glaucoma (high pressure) of the eye and bladder cancer. My risk was supposedly low for obesity, Alzheimer's, Parkinson's or heart disease. This came as some relief, as my father had died young of heart disease. I next looked at the personal traits and found I had an increased likelihood of having curly hair, brown eyes and not being bald, all of which were – so far – true.

I compared the results with those from the other company, Decode, which used very similar genetic methods but a different prediction algorithm. About half of the tests didn't agree well on my risk. For the diseases, they only agreed on an increased risk for adult-onset diabetes, which my grandmother had suffered from. For the personal non-disease traits, according to the Decode results I was 85 per cent likely to have blue eyes. This was a shock, as mine are very dark brown and I'm dark-skinned – so this was one of the few predictions I could show were wrong.

With my scepticism heightened, I looked more carefully at the diabetes results that showed me to be at high risk. In fact, my lifetime risk only increased from an average 15 per cent to 19 per cent – so this extra 4 per cent was unlikely to be life-changing. But the sheer range of other results produced from this single DNA test is impressive. I was found to be at twofold risk of overreacting to blood-thinning drugs like warfarin, but not to a range of other common drugs. This might be useful if I were one day to need this drug, say after getting a blood clot in my leg or lung after sitting for too long on a plane. The results also reassured me that while I was, like most Europeans, able to digest milk and not intolerant to the protein lactose, I had a greater than average chance of being a heroin addict. More worrying than my potential drug habit was me being diagnosed as a carrier of a rare disease I'd never heard of: Canavan's disease, which could affect my children.

As part of the screening test, the 23andMe company included 20 so-called 'monogenic' diseases caused by a mutation in a single gene. There are several thousand of these so-called Mendelian diseases, named after the monk Gregor Mendel, including diseases like sickle-cell or cystic fibrosis. Although most are incredibly rare, they often lead to serious mental, visual, lung or nerve problems and early death. These illnesses make up only 2 per cent of all genetic diseases and are very predictable, which is why they have been portrayed as poster boys for the way all the other 98 per cent of genes and diseases behave. Often a tiny change (mutation) in one single chemical base of a gene or pair of genes could lead to a faulty protein being produced, resulting in the disease.

So I found out I was the proud carrier of this rare and lethal genetic brain syndrome called Canavan's disease, caused by producing an abnormal chemical, aspartate, in the brain. As a carrier I have the mutation only on one of the two copies (alleles) of my genome – the other normal gene means that the enzyme is still being produced normally. It could however be lethal for my children if my wife also was a carrier and had the same mutation on one copy of her genes. Fortunately, although carriers are fairly common (1 in 40) in East European Jews (my father was Jewish, so probably to blame as the likely carrier), the chances of both partners having it are slim (1 in 40 × 1 in 40 = 1 in 1,600), and very much less if one partner is not Jewish. Thousands of other rare genetic diseases act in this way, which is called recessive inheritance.

This real story of personal genetics highlights the successes and limitations of the modern revolution in genetics and personalised medicine. The fact that, only ten years after the first mapping of DNA, any member of the public can now easily access one million genetic tests for a few hundred dollars is pretty incredible. New advances continue at a dizzying pace. Companies like 23andMe are now offering direct to the public for less than $1,500 even more detailed tests using sequencing technology, which can pick up millions of very rare gene variants by only looking at the 1 per cent of your total DNA (your genome) that contains genes. These regions are called your exomes.

Extraordinarily, most doctors and health professionals (unlike the general public, or at least many cab-drivers that I meet) are unaware of these rapid advances and the availability of genetic data. In regular seminars I give to junior doctors, only one in twenty knows how many genes there are in the body – most overestimate by a factor of at least 1,000 and assume there are millions rather than thousands. It just shows how easy it is for the overworked medical professional to be left behind by scientific progress. This gene technology has practical implications and has been great for screening for rare diseases, targeting expensive cancer drugs to most receptive patients and predicting the exact safe dosage of blood-thinning drugs like warfarin.[5] But it also has much to tell us about evolution and where we come from.

We now know exactly when we split from other primates (6 million years ago) and from Neanderthals (half a million years) and that we are much more similar genetically to our Neanderthal cousins than we care to admit – as we can often observe on any Friday night in big cities. We know that South American Indians migrated from Asia across Siberia 5–15 thousand years ago and that Europeans branched off from Asian Indians, who were one of the first modern human groups to leave Africa. So this more precise knowledge of our DNA has altered our understanding of human history.[6]

What about helping us understand diseases? Just in the last few years there have been over a thousand genes discovered that have a role in over 100 common diseases. The early discoveries showed large effects (explaining over 25 per cent of the possible genetic influence) of certain unexpected genes for a few diseases like macular degeneration – the commonest cause of blindness[7] – or for that important public health and personal distress problem – male pattern baldness, which my team helped discover.[8] Having a high risk of both of these 'diseases' can now be predicted with reasonable accuracy from DNA testing. There have been many scientific breakthroughs in just a few years.[9]

However, despite the extensive list of successes, a few signs were emerging that the paradigm was wrong. Most of the gene discoveries for common diseases turned out to be interesting in terms of biology, but the more we discovered the less useful each new gene became in accounting for the disease, since each gene is of tiny individual effect. For example, the 30 or so genes discovered for obesity, even when combined, account for only 2 per cent of the disease.[10]

This was frustrating to all of us working in the field, as it meant that each common disease was controlled not by one gene but by hundreds or even thousands of genes. This would require teams from many countries to combine forces and perform studies of tens, and sometimes hundreds of thousands, of subjects in order to find these tiny effects. Another consequence was that for common diseases (unlike rare monogenic diseases) these gene tests were pretty useless for prediction, as I found out from my Internet results.

Another widely held belief that bit the dust was that only the part of DNA containing the genes was important. The remaining 98 per cent of our DNA was thought to be worthless, containing remnants of old unused genes and boring repeat areas. Yet these non-gene regions are also faithfully copied and inherited, so presumably they once had some use in evolutionary history. The fact that genes and their regions are not all-powerful was highlighted when the estimate for the number of genes we are supposed to possess dropped fourfold from 100,000 to less than 25,000 – pretty much the same number as a worm. For the scientists who believed that genes represented the Book of Life it seemed unlikely that worms and humans were reading the same best-sellers. Even the least sophisticated human we can think of is clearly more complex than a worm.

So our complexity and our many differences cannot now be attributed solely to our genes. The big genetic difference between us and worms is that although we have very similar numbers of genes and gene regions, we have masses of what was until recently and quite mistakenly described as 'junk DNA' – what we now call non-coding (or intronic regions), as opposed to the exonic regions mentioned earlier.

The traditional paradigm of one gene equals one protein and so one disease has also been exposed as actually a rare phenomenon. The same gene can produce hundreds of different proteins via rearrangements (called splicing) due to this 'junk' and to other chemical signals in the cell. So the same gene can produce very different proteins in different environments and therefore different diseases.

The missing 95 per cent

Another gene mystery was how, if every cell in our body was derived from replication of the same fertilised egg and had the same identical DNA, did the original cells manage to differentiate into 200 different cell types as diverse as skin, liver cells and brain cells, each with a completely different function? Until now scientists have ignored this awkward problem, as they did not have the tools to investigate it. Recent discoveries are changing this. The secret could lie in the junk or non-coding DNA carried along in the genes, but this appears to be identical (just like the gene regions) in all DNA throughout every cell in the body.

So something else must be going on that makes cells different – something that can't be driven by the genes themselves. This ability to signal genes to perform different functions and make different tissues has to be coming from the cell itself.

More and more, as we acquire the molecular tools to look more closely, we see greater levels of complexity. Eye colour was until recently thought to be genetically simple, a reliable guide to whether your dad was the milkman; it was believed to be controlled by only three genes. As part of an international research project, my team has shown that it is influenced by at least 20 and possibly hundreds of genes.[11] I have also met a few rare identical twin pairs with different coloured eyes to each other – a phenomenon that was said to be impossible.

While the hundreds of recent gene discoveries have given us great insights into new disease mechanisms and possible drug targets, the common genes found to date usually account only for less than 5 per cent of the genetic influence. Exactly where the missing 95 per cent comes from is a mystery that is perplexing the field. Most scientists agree that we simply aren't yet smart enough to realise what we don't know.

Meanwhile newspapers and the media continue to happily pump out more and more stories proclaiming 'The gene for fat / depression / strokes / homosexuality / anorexia found', assuming a determinism that makes most of us feel instinctively uncomfortable when we think about it. Throughout this book, I will refer to some of these simplistic deterministic ideas and slogans to challenge these outdated newspaper headlines and introduce some more modern concepts.

Generally resistant to the genetics revolution have been another group of scientists: the epidemiologists, those who study the environmental causes of disease, as opposed to genetic epidemiologists like me, who study its genetic causes. Until the last ten years they were the most powerful research group studying disease causes, and attracted most of the funds and glory. The different views of the two groups on the relative importance of genes versus environment accentuated the 'Nature versus Nurture' debate.

The traditional epidemiologists had also made bold claims, such as that 80 per cent of common disease and 30 per cent of cancer in a population is preventable by changing diet, exercise, and controlling smoking and alcohol consumption.[12] In fact, apart from successfully reducing cigarette smoking, which accounts for 30 per cent of cancer deaths, most public health preventions for common diseases have failed. Despite this failure and the costs to society,[13] there has been a scarcity of new insights and ideas in the last 30 years. In particular, little thought has been devoted to explaining exactly how different environments exert their effect and how they can interact with genes.

As a junior doctor in the East End of London in the 1980s I was struck by the number of wheelchair-bound patients I saw suffering from the crippling and deforming joint disease rheumatoid arthritis. They had usually developed this in middle age in the 1950s and 1960s. Nowadays the rate of new cases of the disease has halved and cases are much milder, and in my clinics nowadays I virtually never see anyone in a wheelchair. Doctors like me often take the credit and claim that greater skill, insight, earlier diagnosis, scans and better treatments are the reasons for the change.

Ironically, the reality is that the changes had already started before the new technologies and all the new powerful drugs came into use. Yet the reasons for this change remain unknown. Recent changes in asthma, allergies, short-sightedness, heart disease, diabetes, schizophrenia, autism and many cancers also remain largely unexplained by known environmental factors.

All the usual suspects – alcohol, coffee, tea, sunshine, exercise, diet – get huge media coverage, but usually have tiny individual effects on any disease. In fact, if you put all of the known measurable environmental risk factors together (apart from smoking or age), they can predict or explain less than a twentieth of most diseases. The rest remain a complete mystery.

I have always been interested in solving puzzles: at 14 my career guidance test told me I should be a detective or a psychiatrist (I ruled out being a priest, which sounded less glamorous). The first research paper I published while a medical student showed that consumption of coffee or soya was linked to cancer of the pancreas. I now realise this was almost certainly wrong, and not a cause of the cancer, but the paper did get me hooked on research. Since then epidemiologists have slowly run out of new factors to study, and to further complicate matters it turns out that many of their prime suspects were actually heavily influenced by genes, so not 'purely' environmental at all. We now know that whether you eat garlic or take regular exercise, drink milk or smoke cigarettes is influenced by your genes regardless of your environment. There are few if any examples of environmental factors without a genetic component, and conversely genes don't work alone and are usually dependent on the cells they live in and their environments. So in a world where hundreds of genes are working together to influence a trait or disease, the old distinction between nature and nurture is simply no longer relevant.

To understand which traits are predominantly genetic and which acquired, scientists have, since the 1920s, turned to the simple model of twins. This is a unique 'natural' experiment: the key is to compare the similarities or differences between the pairs. There are two main types of twin: monozygotic (MZ), meaning one-egg, or identical, who share 100 per cent of their genes; and dizygotic (DZ), two-egg, non-identical or fraternal twins, who share on average 50 per cent of their genes and are the same as all ordinary siblings in that respect. In contrast, both sets of twins share very similar environments, so if the factors you are testing (such as height) are similar in both types of twin, a trait can't be very genetic. But if identical twins are more similar than fraternal twins – and given that the other, non-genetic, factors are equally similar – then a trait must be partly genetically controlled. The standard twin study assumes that the family environment is the same for both twins and that the general environment they are exposed to is very similar – which it usually is.

But just how important is the environment, and how do you test it independently? An ideal but cruel experiment would be to obtain two identical clones born at the same time, separate them at birth and give them different environments, and then see what happens. While we are still some way from deliberately creating human clones (and aren't yet allowed to), nature has created its very own unique experiment for us to observe. There are now globally around 11 million natural identical-twins experiments to choose from. From these, a few rare sets have been studied in great detail. They are especially unique as they were separated soon after birth to single mothers by overzealous adoption agencies, particularly in the US and Sweden in the 1940s and 50s, and raised in different families for 'their benefit'. The largest collection of these reared-apart twins features in the Minnesota Twin Study.

Identically different Jims

The tale of the Jim twins is one of the iconic stories that changed perceptions of the Nature versus Nurture debate in the 1980s and helped launch the Minnesota Twin Study.[14] In Southern Ohio in 1979 a 39-year-old called Jim Lewis tracked down his long-lost identical twin brother, who was living locally. They had been separated after their unmarried mother gave them up for adoption as month-old infants in 1940.

On reuniting, the twins discovered that they shared the same first name and an astonishing list of other similarities. They had such a close physical resemblance they both described it as 'like looking into a mirror'. They were, 38 years after they last met, exactly the same height and weight. They had both married and divorced a Linda, and then married a Betty, and had a childhood dog called Toy. They called their first sons James Alan Lewis and James Allen Springer. They both drank Miller Lite, liked beach holidays in Bas-Grille Florida, smoked Salem cigarettes, suffered from migraines and were chronic nail biters. They also wrapped rubber bands around their wrists, hung keys from their belts and used the same rare Swedish toothpaste called Vademecum. At school they had both liked maths and hated spelling. They had carpentry as a hobby, had been part-time sheriffs and drove the same model and colour Chevy.

Sounds spooky? It certainly gives the impression of inevitable determinism, in which environment and upbringing didn't count. Just think of the complex but identical genetic mechanisms that occurred to make them buy the same toothpaste. Going by this example, it doesn't seem to matter what kind of parents you had, and whether you were raised in a hovel or the Hyatt: your genes remain totally dominant. The press loved this story and the two Jims became overnight celebrities and even appeared in *Time* magazine. They were physically so strikingly similar and had so many amazing quirks in common that one of the researchers was quoted as saying: 'this would swing the pendulum even further away from radical environmentalism.'

But could this example possibly have been a combination of coincidence and hype?

They did have some small differences which we now know were downplayed at the time. One had married a third time – the knowledge of which was probably a worry for the other twin's wife. They had different hairstyles – one Jim had a Beatles haircut while the other had long sideburns and a Robert De Niro look. One preferred written communication and the other preferred talking. Other differences would have emerged in the 15,000 questions they answered.

There are a number of reasons their similarities could have been exaggerated. The first is that the environment and peer groups they shared are likely to have been very similar: suburban Ohio is not San Francisco or London in terms of a varied social and cultural mix. Most guys there at some time drank Miller Lite and smoked Salems, did carpentry and had bought a Chevrolet.

The second factor is selective reporting. We all have thousands of personality traits and likes and dislikes that characterise us as individuals. On average we are likely to share many with unrelated strangers from the same country – say our favourite music, food, blend of coffee, newspapers, football teams or TV shows. Some strangers will by chance share more of these. So by picking the most striking pair, and focusing on the similarities while ignoring the traits that they don't share, the similarities are easily exaggerated.

Producers of daytime TV shows are often desperate for twins who say they have psychic powers, and I receive regular requests for help. When we did an actual survey, 66 per cent of our identical twins said they had no psychic powers whatsoever. These 'media boring' twins unsurprisingly never get onto TV, to scupper the idea that most twins are telepathic. Psychologists and advertisers observe that people make subconscious choices. For example, they tend to select mates with similar names to themselves, to name their children and dogs in non-random ways, or to pick products that in some way remind them of themselves.[15]

A final factor in distorting the characteristics of the identical Jims is that once a twin pair has been told they are interesting or similar, they tend to subconsciously exaggerate their similarities and downplay their differences. Although the scientific papers that reported the findings did also mention subtle differences between them, the public and popular impression absorbed via the media was of an uncanny, nearly supernatural resemblance due to the power of genes, leaving little if any room for any other factors.

A new Darwinism?

This book will show why the similarities between the two Jims are the exception and not the rule. To do so we have to look at genes in an entirely new way, and challenge some of the long-held traditional assumptions about our relationship with our genes.

Assumption One is that our genes single-handedly define the essence of human beings: that they are our 'human blueprint' or 'book of life' and are the only mechanism of inheritance. In order to fully understand this point we need to reconsider our entire gene-centric view of life. Assumption Two, following on from the central role of the gene, is that genes and heritable genetic destiny cannot be changed or modified. And therefore Assumption Three is that an environmental event can't produce a long-lasting influence on your genes throughout the cells of your body. Assumption Four is that you cannot inherit the effects of your ancestor's environments – in other words that you cannot inherit acquired characteristics. So the traditional view has always been that, to give one example, the smoking habits of your biological father, if you were adopted, and never met him, could not possibly influence your own health. The new science tells us something very different.

In short, we need to look again at our entire conception of genetic inheritance and question each assumption that has been handed down to us. To do so we will first hark back to a time before Darwin, to consider some alternatives to Darwinism.

1
THE GENE MYTH

Toads, giraffes and fraud

Genes and environment or nature and nurture cannot, as currently portrayed, explain why our genetically identical cells are so different, or the greater-than-expected differences between relatives or twins, or the countless examples of rapidly changing patterns of disease. This brings us to an eighteenth-century scientist called Lamarck and his ideas of 'soft inheritance', which we now call epigenetics. Epigenetics could be the missing third element, alongside nature and nurture.

Jean-Baptiste-Pierre-Antoine de Monet, Chevalier de Lamarck, was born in 1744 near Amiens in northern France. Lamarck was a man of many talents. He started first as a soldier in the war against Prussia, then studied medicine, fossils, then finally botany. He worked in difficult conditions at the time of the Great Terror following the French Revolution, when many of his colleagues had been guillotined for saying things that were not politically acceptable. He published his major works during the Napoleonic wars, and 50 years before Darwin he developed an elaborate theory of evolution.[1]

Lamarck was the first to properly study invertebrate animals, and he was an early champion of the controversial idea that something other than divine intervention was responsible for generational changes in plants and animals. His main theory, formulated in 1809 (the year Darwin was born),[2] was that there was a combination of two evolutionary forces. The first was the vague power of complexity, or *'le pouvoir de la vie'*: simple organisms spontaneously emerge and then slowly evolve to become more complex. The second was force of circumstance, *'l'influence des circonstances'*: animal species had adapted rapidly to their surroundings, and formed habits that exercised and improved (or lost) certain characteristics, such as eyes, tails, colours and muscles. But not only did they seem to be able to adapt to their environment, they also passed some of the newly acquired characteristics on to their offspring. This process has since been called the inheritance of acquired characteristics or 'soft inheritance'.

Like Darwin after him, Lamarck did not use the term gene – the concept was unknown to him and his contemporaries – so he could not explain how these characteristics were passed on. His most quoted example from the many he used is that of giraffes. The tall trees that giraffes ate from, he argued, made them stretch their necks, and the continuous stretching released fluids that made each generation have slightly longer necks. Until recently he and his giraffe neck theory were the butt of many jokes. His observation that plants adapted to different types of soil that they were planted in was more acceptable to his peers – although these possible 'epigenetic' effects on plants were seen as too far removed from 'divinely created' humans to be taken seriously.

Lamarck's theory of evolution was heavily criticised in France by his peers and was soon forgotten. Unlike Darwin, who after a long struggle with creationists ended up triumphantly buried in a prestigious plot in Westminster Abbey, Lamarck finished his life blind and penniless, dumped into an unmarked limepit somewhere in northern France.[3] Even after his death, his French colleagues continued to demean and ridicule him, notably in the so-called 'eulogy' given by his rival George Cuvier in Paris a few years later. History likes winners and losers, and many a schoolchild since then has learned of the foolish Lamarckian theories, trumped by the brilliant and logical Charles Darwin. But the reality was not so simple. Darwin was actually an admirer of Lamarck, and his works contain several references to the notion that inheritance of acquired characteristics might be an alternative or parallel method of evolution, albeit more minor. But at the time most of the scientific world was more interested in our descent from apes and did not listen.

Nazis, communists and Indian ink

The work of the unfortunate Chevalier de Lamarck was not the only precursor of modern epigenetics. Paul Kammerer was a Viennese musician turned biologist who in the 1920s had a fascination with amphibians and with Lamarck's ideas. He was a busy man who like many men of his day performed experiments in diverse areas. His own – and from today's perspective rather eccentric – theories of life were often quoted by Freud and others. Kammerer claimed (without formal proof) that he had skilfully manipulated and bred cave-dwelling salamanders (olms) with no eyes to be able to see. He raised salamanders in very different breeding environments and apparently altered their offspring's breeding patterns.

He was most famous for claiming to have made midwife toads breed in the water as opposed to on land, just by raising water temperatures. The midwife toad gets its name because the male carries the fertilised eggs around on its hind legs. He also reported that his new generation of toads were now exhibiting black nuptial pads on their feet with tiny spines to stop them slipping during mating in water, just like their distant ancestors.

Kammerer drew big audiences for his international speaking tours, which made him good money. The *New York Times* in 1923 hailed him as the new Darwin, having proven Lamarck's ideas of inherited acquired characteristics.[4] He even had a celebrity mistress, Alma Mahler, the newly widowed spouse of the late Gustav Mahler. Alma Mahler was the femme fatale of her time, who while picking her way through famous musicians and artists, also worked as Kammerer's assistant. She complained about his sloppy record-keeping and over-eagerness for positive results. Kammerer soon became known in Vienna as the 'Wizard of Lizards', as much for his wild social life and strong socialist and pacifist views as for his science. He had also irritated some Americans both from his hyped success in the media and because when he visited America during Prohibition he predicted piously that future generations would benefit from the alcohol-free environment of their parents.

But Kammerer's fame was not to last. Scandal hit when in 1926 the journal *Nature* published a letter stating that the famous toad experiment had been faked. G. K. Noble, Curator of Reptiles at the American Museum of Natural History, had visited his old lab in Vienna unannounced when Kammerer was still on his money-making world lecture tour and inspected the famous specimen of the preserved but long-dead toad. The black pads, Noble claimed, had a far more mundane explanation: 'it had simply been injected there with Indian ink'.[5] Six weeks later Kammerer shot himself in the forest of Schneeberg, leaving a suicide note with a somewhat ambiguous content. 'Who besides myself had any interest in perpetrating such falsifications can only be very dimly suspected,' he wrote. This note was also, strangely, published in *Science* – an unorthodox posthumous way of improving your CV.[6]

Interest in Kammerer's experiments revived 40 years later in

1971 with the publication of a book on the incident by the Hungarian author Arthur Koestler. In *The Case of the Midwife Toad* he suggested that the toad experiments might have been doctored by an early Nazi sympathiser (a so-called Hakenkreuzer, swastika-lover) at the University of Vienna where political activism was rife.[7] Koestler also pointed out that the dodgy toad had been exhibited earlier in 1923 in Cambridge to known sceptics who had examined the specimens and hadn't spotted the crude ink injections and claimed to have seen the spines. This suggested that the ink could have been added later.

In 2009 a Chilean biologist, Alexander Vargas, reignited the debate by elevating the vilified Kammerer to the status of the real father of epigenetics and Lamarckian biology. He examined Kammerer's lab books and breeding experiments, and concluded that many of his findings that were ridiculed in the past could now be supported by modern science and our understanding of so-called imprinted genes.[8] Not everyone agreed. A subsequent editorial and some detective work in an American biology journal in 2010 showed evidence that he had a track record prior to the toad incident.[9] He had previously tried to artificially touch up an image of a salamander's spots while submitting an article for the same journal. They damned him a second time as a fraud and a bad example to others. However, they also admitted that even today, up to 25 per cent of scientific images submitted to journals have some degree of 'enhancement'. So whether the Wizard of Lizards was just a confident fraudster or a genius who was the first to show Lamarckian inheritance, as well as a victim of jealous Americans and Nazi saboteurs, will never be known for sure.

Two years later in 1928 another remarkable, if unpleasant, scientific character emerged from Stalin's Russia. Trofim Lysenko may have unwittingly cost Russia the Cold War 50 years later.

He was a Ukrainian self-taught biologist of peasant stock who embraced neo-Lamarckism. Like Stalin, he disliked the Western- and then Fascist Germany-dominated world of traditional genetics run by elite intellectuals.[10] Their ideas of genetic determinism, eugenics and the power of heredity ran against socialist ideals, which rejected inherited privilege.

Lysenko first came to Stalin's notice by performing an amazing farm experiment. This happened during the new collectivisation policy of changing small family-run farms into state cooperatives. Local Soviet methods to improve agricultural output were given top priority as part of the new five-year plan. Lysenko took one large farm's entire seed supply (without their approval), wetted the seeds and buried them in sacks in the frozen ground to 'prime' them for the next year's harvest, so that they and their progeny would be tougher and produce more wheat. The results were spectacular and more experiments were started immediately, slightly altering the conditions of the priming, or vernalisation, as it was known. Stalin loved the simplicity of his approach, and its PR spinoffs, as all peasants could now become barefoot scientists as well as farmers.

With no need to rely on the infrastructure provided by universities, and on complex and expensive lengthy plant-breeding experiments, Lysenko offered immediate solutions to Stalin and rapidly gained power and influence, becoming head of Soviet biology. He entertained visits from prominent US and European scientists eager to understand his vernalisation methods. But there was a dark side. Anyone who challenged his unorthodox methods or results, or openly supported Mendel or Darwin, was viewed as a traitor to the revolution and either shot or sent on permanent sabbatical to the Gulags. In 1948 genetics was officially banned; it was called a 'bourgeois pseudoscience' until 1964.

There was, however, one little problem with the Lysenko alternative of Lamarckism. It was all one big lie. None of his experiments ever succeeded. No crop yields increased, no trees grew. Failures were covered up, although the rolling programme continued to obscure the truth. Millions of Russian peasants died of starvation, and because of the long-term lack of scientists and plant-breeding innovations, postwar Soviet Russia embarrassingly ended up dependent on America for its food imports. The USA had meanwhile successfully bred maize hybrids using traditional Mendelian genetics and were now tripling their yields. The collapse of the Soviet empire was not due to its failures in arms or technology, but ultimately to failures in agricultural genetics and biology.

But the man who, in retrospect, can be regarded as the real father of modern epigenetics was Conrad Waddington, an Englishman born in India in 1905, who was way ahead of his time. He too started in science with a strange interest in amphibians and how they developed – though he wisely stayed clear of toads. He moved on to study genes and heredity in fruit flies. He was, just before the Second World War, the first to suggest and use the term epigenetics, derived from the Greek prefix epi-, above or around, and genetics. He was fascinated in early development of the fetus and interested in the mystery of how cells can start so simply and then develop specialised functions, yet all have the same genetic material.

Before the structure of DNA was discovered, Waddington believed that tiny changes (mutations) around our genes could lead to differences in the way that cells and whole animals develop and could in theory be passed down generations. As a Fellow of the Royal Society, he was one of the most eminent pre-molecular developmental biologists of his time and his work suggested that some of what Lamarck had said might just be correct.[11] Unfortunately, following the stir caused by the elucidation of DNA structure and the molecular biology of genes, his work was overshadowed and forgotten for many years.

How do plants know when to flower?

Looked at from today's perspective, how does Waddington's or indeed Lamarck's theory hold up?[12]

While Lamarck made some very interesting and relevant observations, he should perhaps have steered clear of talking about both giraffes and lettuce in the same breath. Plants and animals differ in quite a few ways. One difference is that plant cells are pluripotent (multipurpose): they can all change to another form if needed and become specialised. In this way small cuttings can sprout a whole new plant – unlike someone attempting to plant a human finger. This means that they must have ways of modifying the genetic information from the identical DNA contained in each cell to provide the message to make a specialised daughter cell. Epigenetic mechanisms were supposed to play a part in this, but after the cells divide, these new signals were believed to be wiped clean again, so that the cell could remain pluripotent. This would mean that cells had no remnants of interfering messages – for example trying to make cells become leaves or roots. The idea that all memories of how a cell had diversified were completely wiped clean as the pollen (sperm) and the egg (called gametes) were formed to make a new generation has been central to the traditional view of genetics. We now know this isn't exactly true. The wiping process isn't perfect.

Over ten years ago a group in Norwich discovered a natural case of epigenetic changes in plants. Remember this means a heritable effect that is not due to changes in DNA structure. A 'mutant' version of the common toadflax plant, a pretty yellow wildflower growing in hedgerows, results in flowers with radial petals (five), rather than the normal two.[13] What was unusual was that although the DNA structure was the same in both plants, the 'mutation' could still be passed on. Normally a mutation is a change in the actual DNA – which was not the case here. The researchers found this change was due to something called 'methylation', which is a key part of epigenetics in animals as well as plants, and one that we will return to later. In the mutant plant, a key gene (called Lcyc) is extensively methylated and in the normal plant it is not.

What methylation means is that at certain sites (usually cytosine bases) of the gene's DNA, small chemical methyl groups (Me) floating around the cell attach themselves to it, rather like sticking an olive on a cucumber with a cocktail stick. This has the effect of stopping the gene producing a

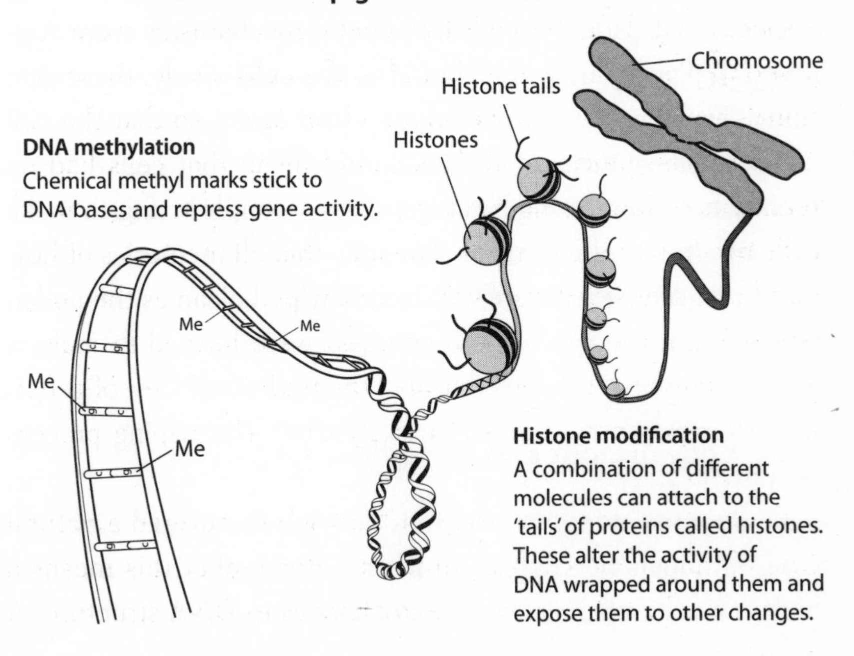
The two main epigenetic mechanisms
Chromosome
Histone tails
Histones
DNA methylation
Chemical methyl marks stick to
DNA bases and repress gene activity.
Me
Me
Me
Me
Histone modification
A combination of different
molecules can attach to the
'tails' of proteins called histones.
These alter the activity of
DNA wrapped around them and
expose them to other changes.

protein. We call this inactivating it or 'switching off', and we know that in most cases methylation stops a gene from working, or 'being expressed', while reversing the process (un-methylating) usually switches the gene back on. By being turned on we mean that it is expressed and more protein is produced. While this process, unlike a mutation, is reversible, it can also last a long time.

The Norwich team found that most subsequent generations of toadflax plants had the same radial petal pattern, and carried the same deactivated gene due to methylation. This showed that the information couldn't have been wiped and reset as previously believed. This was the first clear modern evidence that natural epigenetics occurs and can be passed on across generations. Others were soon to follow.

How do plants know when to flower? It seems a simple question, but until recently we had no idea of the answer. The arabidopsis plant (thale cress) alters the timing of its flowering by epigenetics.[14] In response to prolonged cold (as in winter), the Flowering Locus C gene which normally prevents flowering is methylated and deactivated, allowing this variety to flower in the spring. The trait is then passed on to the next generation, even if there is no cold winter. Ironically this experiment showed that the vernalisation mechanism favoured – and faked – by Lysenko was actually a real biological phenomenon.[15] If he had actually tried to do proper experiments, and politics and science hadn't clashed, he might have produced valid results.

Fleas, body armour and butterflies

Plants are one thing, but animals are obviously more relevant to us. Although we share around 40 per cent of our genes with the banana, we share more genes and genetic mechanisms with other animals. So can genes also be modified in animals, and can they indeed be passed on to the next generation as Kammerer described for the foot pads of toads?

The water flea (Daphnia) is a tiny aquatic species that as adults have a variety of defences against predators. These include helmets and spiny tails. Some fleas have both, some have one, and some cool fleas none at all. What is strange is that these fleas have identical DNA – like identical twins.[16] Place a young flea in water with no trace or odours of predators and they will develop no defences. But if you put its genetic clone in another tank with old traces of a nasty fish, it will develop a spine and helmet. Put the babies of these two in opposite environments, and they will be armed according to the environment of the mum rather than their current aquarium. Just as intriguing is the fact that this effect lasts a few generations and fades.

Caterpillars and butterflies look and behave very differently but they have exactly the same DNA structure. Their cells have developed differently, and these differences must therefore be epigenetic. Butterflies have also been found to show very different mating tendencies if the temperature on the day they were born varies by just a few degrees. Female squinting bush brown butterflies who developed as caterpillars in cooler temperatures (17°C) were more likely to have flashy wings and be chasing males than their genetically identical twins brought up at 27°C, who behaved more demurely.[17]

We don't normally give much thought to the billions of chickens bred each year, unless it's to how we like them cooked. But they can be very useful for research. Chickens make ideal adoption experiments, as their eggs can be nurtured and hatched and the chick reared with no contact from either parent or social workers. In one experiment parent chickens from the same genetic stock were raised in one of two scenarios: one was a comfortable private clinic-style environment of 12 hours of daylight then 12 hours of night (called a predictable light rhythm) in which they could eat in a relaxed way. The other was a Guantanamo Bay-style scenario where they had sadistic unpredictable light rhythms which, as they only eat in daylight, meant they had unpredictable eating opportunities that could be halted at any moment.

The researchers then looked at the eating patterns of the offspring of both groups when they were now all raised uniformly in the comfortable regular daylight and eating lifestyle. Those chicks from parents raised in the Guantanamo Bay-style who had never met either of their parents had a more efficient and aggressive eating behaviour than their genetically identical but parentally privileged coop mates, who were more relaxed and preferred to look around for more tasty worms who often got away. The efficient policy worked and the Guantanamo chickens got fatter. The researchers saw epigenetic changes in the offspring and suggested these had affected immune and hormonal genes such as oestrogen.[18] This suggested how epigenetics could provide survival advantages. Environmental stresses could prime future generations to be able to cope better in the same situations – a brilliantly effective form of short-term evolution.

So what about animals that are more similar to us than butterflies and chickens? Mammals like mice share around 90 per cent of their genes with us, including most of the known disease genes. A remarkable experiment by Randy Jirtle at Duke University[19] has shown that simply by slightly altering the diet of a certain type of pregnant mouse (called Agouti mice) you can change their offspring from chubby blondes to skinny brunettes; and moreover that this effect of grandma mouse's diet can be passed on for three or more generations until it fades. The Agouti gene normally produces a yellow fur pigment, but if it is switched off by methylation-inducing chemicals in food, it produces a brown pigment. This reversible inherited change, which does not alter the DNA structure, is the essence of epigenetics.

Until recently it was thought that these epigenetic findings were just in a few rare or exceptional genes – so-called imprinted genes – which are much more common in mice than man. Remember we inherit two copies (alleles) of every gene: one from each parent. In most animals a major battle goes on between the genes of the usually absent father and those of the mother. The father's genes are trying to increase the size of the fetus – so that it has a greater chance of survival – at the expense of the mother,[20] who is trying to conserve her resources and live long enough to have more children. In mice, the mother usually wins: in several hundred genes she manages to permanently suppress the father's copy of the gene by this imprinting mechanism and so keep the fetus a manageable size. Humans have around 50 of these imprinted genes, which form the battleground of parental gene warfare and have a major role in the size and development of the fetus. While these 50 genes are important,[21] we know now that the rest of our 25,000 genes can also be influenced epigenetically.

Soft inheritance: nature vs nurture revisited

These recent exciting findings in animals have confirmed work and ideas dating back to Lamarck, suggesting there is more to the inheritance of genes than just the painfully slow process of Darwinian evolution and natural selection. Soft inheritance is the parallel faster route by which we human beings adapt to our surroundings, and also explains many of the emerging ideas of how we are moulded into individuals.

But before we get into the extraordinary implications of this new understanding of soft inheritance, it is worth considering how our attitudes to traditional inheritance have altered over the last 50 years.

'Happiness gene discovered' leads the headline from the UK *Daily Telegraph* in May 2011.[22] 'Those with two sets of the gene – one from each parent – are almost twice as likely to say they are satisfied with life, compared to those who lack a copy'. We are now becoming blasé hearing about these stories over our daily cornflakes or muesli: the media have to sex them up to grab our attention. Nearly every disease or behaviour studied has shown some influence of our genes – and most human studies involve twins. Many of these twin studies have totally changed our perceptions of diseases. These include 'boring' wear-and-tear diseases of old people such as arthritis of the knees, back pain, cataracts of the eyes,[23] varicose veins or even haemorrhoids.[24] All of these turned out to have surprising major genetic influences.

Other real breakthroughs have been achieved in areas of personality and behaviour such as autism and schizophrenia, where gene discoveries have shifted perception of guilt. Now we accept the involvement of neurochemicals and genes in the brain, rather than just bad parenting and dangerous vaccinations. Genetic influence has been competing with a strong and long-standing Christian culture of believing that sin causes sickness – particularly when the sickness is mysterious or affects the brain. The popular acceptance of twin studies and genetic influence on our personalities and traits has waxed and waned since the 1920s, when the first proper twin studies were started in Germany and the USA.[25] Until the 1950s most people seemed happy to believe in some degree of genetic determinism, and then in the 1960s and 1970s came an environmentalist backlash, often linked to socialist ideals in education and equal opportunities. This look at history reminds us that our views and interpretation of biology and science are just as influenced today by social, religious and political pressures as they were in the careers of Lamarck, Darwin or even Lysenko. We often still believe what we want to or think we ought to believe.

Early twin studies in the 1960s were heavily criticised, usually by social scientists philosophically opposed to the notion of genetic influence on personality and IQ. They said that parents might be more likely to treat identical twin pairs more similarly than would parents of fraternal or non-identical twins. Although difficult to prove conclusively one way or another, there is no clear evidence of a bias by parents. More important, even if there were, its effect would be trivial and would only change heritability estimates by a few per cent.[26] Some of these scatter-gun attacks in the pro-environment culture of 40 years ago did however eventually hit home, casting doubt on some of the genetic findings.[27]

As a bonus, critics gleefully suggested that some twin researchers were fraudulent. An eminent British educational psychologist, Sir Cyril Burt, produced papers showing that IQ was strongly heritable. He was an honorary president of Mensa, had been a member of the Eugenics Society, and helped start the 11-plus exam in schools – none of which were popular in the 1960s and 70s. When he died in 1971 one of his neo-Marxist critics, Leon Kamin, a member of the group Psychologists for Social Action and a junior academic in rat psychology, began his campaign. He found that Burt's original research records and notes had been destroyed. He was suspicious and dug some more, culminating in claims of fraud in a book that was later publicised in 1976 by the journalist Oliver Gillie in the *Sunday Times*.[28] By comparing his published papers they found that Burt unerringly produced the same precise estimates of similarity, even when the numbers of twins in his studies subsequently increased. Although possible, repeating statistics so exactly in science is unlikely. They also accused him of inventing non-existent co-authors. Burt, being dead, could not defend himself, although it turned out years later that the co-authors probably did exist and, with a witch hunt going on, were keeping their heads down. Burt was branded a fraud and the environmentalists used this to condemn all twin and genetic findings.[29] Indeed Kamin, whose career then took off, went so far as to claim that IQ had zero heritability. The power of the environment was briefly supreme again, and genetics squashed for nearly another two decades.

Ironically, Burt's estimates of IQ heritability of 60–80 per cent have since been replicated multiple times in over 10,000 subjects in twin, adoption and family studies, and the idea that he was falsely accused has gained momentum since the late 1980s.[30] There is an interesting historical parallel with allegations of fraud against other eminent scientists, not only Kammerer. Isaac Newton has been accused of rounding up his speed-of-sound calculations and of plagiarism about his gravity theory. In the genetics field, the famous monk Gregor Mendel was the first to describe the concept of heredity and dominant and recessive genes. When counting his peas he managed to round the numbers up very neatly so that they met the expected ratios exactly. But as his theory, like Newton's, proved to be correct, he got away with it.[31] The environmentalist critics of Burt and genetics were vociferous and persistent but were never able to show that twin studies were wrong, just that they had possible flaws.

One of the reasons I started to study twins and set up the

UK Twin Registry in 1992 is ironically thanks to the same pro-environment lobby that destroyed Burt. In the 1970s and 80s in the UK the grant bodies and academics in psychology and sociology were firmly in the pro-environment camp with its dogma that all humans were created equal and that genetic research on IQ was akin to racism. Because of the pressure from the socio-environmental lobby, twin funding dried up for the UK academics. Believing they couldn't beat the tide, they and their gifted teams sensibly left for well funded posts in the US and Australia. This left no large-scale twin research in the UK, where most of the methods since Galton had been invented. In 1992 I was lucky to find the tide beginning to turn. The media and some grant bodies were now generally helpful, and importantly there were thousands of willing volunteer twins keen to help research without a political agenda.

Irritatingly for the environmentalists, although they rightly claimed that some of the anecdotes were exaggerated, they couldn't explain away the overall data of identical twins raised apart. Tom Bouchard and his colleagues in Minnesota in the 1980s had slowly built up an impressive collection.[32] As well as confirming the heritability results from the larger twin studies for diseases like diabetes and obesity, they confirmed an IQ heritability of 70 per cent – similar to Cyril Burt's findings. Their novel observations into our behavioural traits and personality raised many new questions about the nature of our identity and even free will.

In my office at work is a giant poster of hundreds of photos of twin faces that I look at every day. As they pose for the camera, strikingly, all pairs have the same look. Some glum, some shy, some smiley – but all the same. From my interviews with the many twins in this book and over the last twenty years, it has become clear that identical twins, whether raised together or reared apart in humdrum or crazy families, tend to look alike, talk alike and have very similar mannerisms and facial expressions. These are traits which as humans we are extra-sensitive to picking up. These particular traits appear pretty much hard-wired, as I've yet to see any exceptions. However, from examples I use later it is clear that under the surface veneer, fundamental differences often emerge in terms of behaviour and disease.

Changes in behaviour, for example, are driven by changes in our brain – in the billions of brain cells that communicate constantly with each other via trillions of electrical neuronal connections that control all our thoughts and actions. But each nerve cell, which is itself a very complex machine, is driven by exactly the same structural DNA to function via its proteins. These should be exactly the same in each cell in the body and the same in identical twins. But we know that differences do occur, so something must happen to the DNA to make it act differently to produce different chemical and electrical signals that alter our behaviour. Let us start to dissect ourselves and our brains in depth by first taking a look at a unique human trait: happiness.

2

'THE HAPPINESS GENE'

Mindsets, optimism, laughter

I met Barbara and Daphne about ten years ago when they visited my research unit at St Thomas' Hospital in London and we appeared on a TV programme together. They were then in their sixties, having been adopted by different families soon after birth. They became part of the famous Minnesota study of adopted twins that had been reared apart. Barbara and Daphne's mother was a Finnish au-pair who put the girls up for adoption after giving birth to them in London's Hammersmith Hospital in 1939. As a single home couldn't be found for them they were separated. The fact of their adoption was kept a strict secret in both families until the girls started work and needed official birth records.

Daphne was adopted by a metallurgist and his wife in Luton. She went horse-riding and had piano and ballet lessons. Barbara was adopted by a park attendant and his second wife in West Kensington in London, who both died by the time she was a teenager, and was then raised by a strict aunt. Both twins recall being generally happy in their childhoods despite the different surroundings. They were reunited in 1979, when Barbara found out first that their mother had committed suicide, aged just 24, soon after returning to Finland, and then that their mother had also been adopted – her father was a Russian.

'I met Daphne at London's King's Cross station,' says Barbara. 'We didn't hug and we didn't kiss. There was no need.

It was like meeting an old friend. We just walked off chatting together. The funny thing is we were both wearing a beige dress and brown jacket.' It was the first of several coincidences they discovered as they caught up with each other's lives. These included leaving school at 14, having blue wedding themes, having three or more kids and one miscarriage, liking cool black coffee, Marmite, and making the same silly spelling mistakes in tests such as 'the cas [instead of cat] sat on the mat'. They also shared a dislike for answering questionnaires. After answering 15,000 questions for the Minnesota study they vowed never to fill in any more, despite my attempts to persuade them, but luckily they were more than happy to chat.

They had both been raised by rather dour adopted families without much tradition of jokes or humour. 'Neither of us felt we really belonged before we met each other – we had never met anyone who laughed as easily as we did,' Barbara said. 'We now know we have a similar sense of humour but don't particularly like slapstick comedy and are both useless at remembering or telling jokes. We do feel sorry for anyone going out with us, as we don't stop talking and laughing.'

'The giggle gene'

Happiness, or contentedness, is not easily measured or defined, yet we are obsessed with trying to understand how to achieve it. There are over 12,000 books in English on Amazon with 'happiness' in the title, most of them claiming they can let you into the secret of how to attain it – if you can afford the cover price. In the same vein, 'the happiness gene' keeps popping up in the media, with multiple claims to have located it in the last five years, though mostly these are just rehashed stories. There is no single gene that controls happiness, but if there was, Barbara and Daphne, 'the giggle twins', would both have it.

Laughing is part, more or less, of all of our personalities. It is a very basic human reflex that may be related to language and starts as early as 17 weeks of age – according to laugh experts (gelotologists). The 'giggle twins' appear to have inherited an inherent low threshold for laughter as well as an outward expression of happiness that couldn't be dampened by their adopted families.[1]

Studies of 29 other twin pairs reared apart have shown that the particular way they smile to positive emotions is genetically influenced. But strangely the way they grimace at bad news or negative images is not.[2] This suggests that smiling and laughter has its own private brain network and may be more important in our development and evolution than we thought. Detecting angry or sad faces may still be important, and perhaps as it is more crucial to our survival it is not as variable.[3]

Happiness, like humour, is a part of our personalities that is very difficult to define. I was raised listening to my father's witty quips and self-mocking irony, and according to friends have inherited some of his traits. I find my son has also inherited a similar style and even when still young could entertain a crowd. But is this a genetic effect or more the result of myself (and then my son) being exposed to the equivalent of thousands of hours of badly performed Woody Allen and Groucho Marx-type gags? Although it has proved impossible to get funds for, and makes most of my colleagues question my sanity, I have been trying over the last 15 years to find the secret of 'humour' through my research with twins.

To test the ability of hundreds of twins to tell and laugh at jokes and to record them accurately is a tough job and, like many attempts to measure personality traits, tends to produce large errors that mask any genetic component. I had thought of hiring a comedy club with a twin 'open mike' night and twin judges, but there were a few technical issues to solve. First, having to get twins independently to choose the gags; then telling the same jokes separately while the other twin was in a soundproof booth; and finally having the separate twin judges marking reliably without conferring. We also had to mix them around so they didn't get bored of hearing the same jokes repeated by the second twin. I realised this would have been either a glorious disaster or a hit daytime TV show. Sadly we shall never know.

Back in the real world, the first experiment we tried was to look at humour appreciation using visual cartoons and a questionnaire with a ten-point scale. A score of 10 was rated 'one of the funniest jokes I have ever encountered'; a score of 1 was 'not worth the paper it was written on'. As we had to use novel material (for the twins), we picked some of the older, less well known of Gary Larson's *Far Side* cartoons. We selected five in the end that got the most variable response from my team. For a study exploring genetic differences between people there is no point in picking jokes that either everyone or nobody finds funny. It was the differences between people – in science-speak, the variation – that we were most interested in.

We looked at 127 female twin pairs aged between 40 and 50. Overall their ratings for the jokes were not high: most didn't like them. For example, for the hare and tortoise joke (which I liked) they scored on average 4/10 and the identical twins had only a 40 per cent agreement in scores compared with 60 per cent in the non-identicals. When we looked at all five jokes, both types of twin had similar levels of agreement, so our conclusion was that genetic influences were not a major factor. This means that unlike most other traits – and surprisingly to us – it appeared that cartoon appreciation is mainly due to environmental (meaning social, cultural and lifestyle-related) effects.

Another possible reason for our unexpected result was that humour is not an easy trait to pin down experimentally. It could be that we had picked a style of comedy that was just too quirky to study. For those who don't know them, Gary Larson's cartoons are full of paranoia, irony and nature and often have a dark or black humour behind them, such as the hare and tortoise being squashed by a truck, or the deer with a birthmark shaped like a target on his chest, or the kid pushing the Midvale School for the Gifted door with a sign that says pull. Although Larson has fans worldwide and has sold over 45 million books, many people fail to be amused. It could be that for this particular humour style culture really was crucial.

The press picked up on this story in a number of different ways, my favourite being the UK's *Daily Star*, a cheap popular tabloid that has lots of sport and semi-naked ladies but is famous for its catchy headlines and xenophobia. They claimed an 'exclusive' interview with me on page 3 – next to a comely naked lady – with the headline 'Why Brits get Fritz in Fits'. Underneath was a brief explanation that we (the boffins) had found the magic cultural ingredient of British humour – and how this meant we clearly had a better sense of humour than the Germans. A masterful, if erroneous, interpretation of the data.

Although I believed our data was correct, I could not overcome some lingering doubts over our conclusions. Perhaps it was our choice of wacky humour or cartoons that was not representative. All aspects of life that can be measured can be studied scientifically, even jokes. Hans Eysenck, a famous pioneering British behavioural psychologist and student of Cyril Burt, also studied humour in the dark days of 1942, and came up with the now accepted concept that every joke has three possible elements.[4] First, the cognitive aspect: this is the punchline, the 'get it' part, the unexpected surprise twists that make people laugh. The second is the conative aspect – the feeling of superiority at another's misfortune (e.g. what do you call a man with no arms and no legs in a water barrel? . . . Bob). The third is the orective – the sexual innuendo or dirty joke. The best and most successful jokes have all three elements.

The original cartoons that we chose for our first study may have focused too much on the first aspect, the punchline. So we decided to repeat our experiments using a wider variety of jokes, including some the twins had to read out loud to themselves as well as some smutty and misfortune jokes. Our results were not clear-cut. The responses to some jokes did have a clear genetic component but those to other quite similar jokes did not, and were clearly cultural. We tried hard, but still couldn't tease out any consistent patterns in the three joke categories. The secret of the perfect joke seemed to be just beyond our grasp.

A few years later another group of intrepid Canadian joke researchers approached us to carry on our work and answer the question the humour world was waiting for. Importantly, they had funding. They believed that an individual's responses to jokes and humour were too difficult to assess, so instead they used an overall self-rating of humour. Most people when asked a simple question will respond that they have a good or better-than-average sense of humour, so they had to use a more detailed set of questions in a larger number of twins, breaking humour down into four styles and 32 statements.[5]

Two styles of humour were potentially positive assets: feeling part of a group ('I laugh and joke a lot with my friends') and self-enhancing humour ('If I am feeling depressed, I can usually cheer myself up with humour'). Two were negative:

aggressive ('If I don't like someone, I often use humour or teasing to put them down') and self-defeating ('I let people laugh at me or make fun at my expense more than I should'). Using 4,000 UK twins, it turned out that all the four styles of humour were twice as similar in identical as in non-identical twins (meaning that identical twins were twice as likely to score the same in each category as non-identical twins), with an estimated heritability of 40 per cent, showing a genetic component but no clear effect of family environment.

The Canadians simultaneously performed the same study using over 300 US pairs from the Twinsburg annual twin festival in Ohio. This brought surprisingly different results. While both countries agreed on positive humour styles being similarly genetic and environmental, in Americans (unlike the Brits) no hard-wired genetic influences were seen for the negative humour styles. In the US twins these negative humour styles were subsequently associated with a lack of mental toughness.[6] Recent study has shown that Australian twins resemble Brits in humour styles and are also different to Americans. This is a good example of how genetic influences (heritability) can vary between populations. So could evolutionary gene differences or variations explain the greater use and acceptance of sarcasm and self-deprecating humour in weak-willed wimpy Britons? This is improbable, whereas a much greater exposure to 'black' self-deprecating humour over many years and generations seems more likely.

TV shows from the UK such as *Blackadder* and *Little Britain* are famous for their black dark humour. In the TV show *The Office* the unpleasantness of the show's hero, David Brent, had to be toned down for American audiences in the stateside version. So it appears that while most of us have a sense of humour that is both culturally and genetically determined, the ways in which we use or misuse humour are much more variable across countries and cultures, showing that perhaps the *Daily Star* reporter had correctly worked it all out ten years before: Germans just don't have the same sense of humour as the Brits.

'The smell of a new car'

'Overall would you regard yourself as happier and more content than the average person?' Despite life traumas and bad times, over 85 per cent of us rate ourselves as above average in happiness compared to others. This is odd, since clearly most of us can't be above average. It suggests that personal optimism and confidence in the future is another hard-wired human trait. Surveys like a UK MORI 2007 poll show that we believe that improving the following five factors will make us happier: health, family, friends, travel and wealth. But what drives us to believe in these factors without much evidence they are true? Is for example the search for material wealth a natural instinct, or something that recent cultural and advertising trends have fostered in us?

In 1972 the average male was exposed annually to only around two hundred advertisements. Forty years later this had risen to well over three thousand[7] – so many that it is hard to recognise or remember them. Over the last 50 years of advertising the message has remained: 'Buy our product and it will make you happier, more attractive and more successful.' As Don Draper, the protagonist in the TV series *Mad Men*, set in a 1960s advertising agency, memorably puts it: 'Advertising is based on one thing: happiness. And you know what happiness is? Happiness is the smell of a new car . . .' Most experts agree that materialism in society has increased, but was this human instinct always there? The study of 480 US and Canadian adult twins showed that while happiness was consistent with previous studies as being 46 per cent heritable and 52 per cent environmental, materialism was not at all hard-wired.[8] It was not genetic but was entirely influenced by family and the outside environment. So we can blame Don Draper and the Mad Men – not our genes – for this more recent human attribute.

Contrary to our materialistic views, obtaining more money doesn't actually make you happier unless you are very poor. While moving from being poor to average does have a large impact on people's lives, moving from middle- to high-income ranges delivers no significant increase in happiness. Similarly marriage has only a marginal effect on somebody's overall happiness (and it wears off quickly). Health does seem to be important, but only as you get past 65. Summarising the many surveys of what makes people happy, some unlikely candidates emerge: having a higher degree, being involved in religious activities, and gardening.[9] So even pulling a few weeds up helps us somehow find happiness, as we get dirty and back to our roots – ideally also thinking about God and how much fun it was being a lazy student.

Across cultures, people rate subjective well-being as the most important element of their life and even more important than material success.[10] In the regular global happiness surveys so beloved by the media, some countries consistently top the polls.[11] In a Gallup world happiness survey of 2011, Nigeria actually came top, followed by Afghanistan and many Latin American countries, with the ex-communist countries Russia and Romania the unhappiest, and the US and UK slightly higher.[12] In a more detailed study of 15 European countries the winner was Denmark – the UK scored a mediocre ninth. The most miserable complainers were those living in the sunny south

– Greece, Portugal and Italy – and that was before the Euro Crisis. What is so special about the Danes? Not that they have the best weather, jobs or the most money – they don't. It's that they have good social cohesion and the lowest expectations. Similarly in a 2009 US National survey Louisiana (pre-Hurricane) came top and over-expectant New Yorkers came bottom. It appears the less you wish for (and the less you strive for happiness) the happier you are likely to be – suggesting that wild over-optimism doesn't always lead to happiness.

Daniel and Simon are identical twins who led a normal, fairly happy childhood, although their parents divorced when they were teenagers. Simon was always taller, more competitive and slightly better at sports and schoolwork than Daniel. Both twins liked the outdoor and sporty life, enjoying rugby as well as building and DIY. They also remember constantly fighting as boys. They both joined the police force and while Simon stayed and did well, Dan couldn't stand the paperwork and rigidity of the system and went to work on a farm, which he loved. Both twins married and had children and saw each other every few months. Then their lives diverged along very different paths.

When he was 23, Dan was loading hay onto a lorry when he was hit by a 50-kilo bale thrown by another worker. It broke his spine and pelvis in several places. He was in hospital for a year, and it took five years and many operations before he could walk again. He lost his job, his wife walked out on him and he lost contact with his family. His pain has never disappeared completely and he can't run or sit or stand for any length of time. He hasn't had a steady job since and knows he can never play sports again. Understandably he has had spells of depression.

His brother Simon meanwhile rose to become a successful detective in the police. He had a traditional family life and was very content. I spoke to them when they were 55 and asked whether they would have swapped lives. Dan said emphatically: 'No – although I went through a bad patch, I am happy the way things turned out. Although often poor and living on meagre benefits, it allowed me the freedom to travel and see the world. I spent most of the last twenty years in Africa and Asia doing mostly voluntary work with people much worse off than me.' He also met his new wife in Tanzania, and they have bought a 250-acre farm there together, where he hopes to spend the rest of his life.

His successful brother Simon has the nice car, large detached house, golf club, police pension and material success. 'I think my brother and I are so different in personality and lifestyle that we must be non-identical. We really have nothing in common.' However the DNA confirmed their status as clones. 'Although I certainly wouldn't swap lives with Daniel, I do sometimes envy his lack of responsibility and freedom.'

David Lykken and colleagues from Minnesota looked at over 2,700 middle-aged US twins and used a wide range of self-rating statements to measure happiness, such as 'I am just naturally cheerful' and 'My future looks very bright to me'. Results were similar even in a subgroup raised apart and show a heritability of between 40 and 50 per cent. However, happiness is also a transient state of mind that changes with both time and circumstance, so a single snapshot may be meaningless. What the clever Minnesota researchers did was to measure happiness at two time points about eight years apart.[13] This showed that identical twins strongly agreed across time, so about 80 per cent of long-term happiness is genetically controlled – more than at any single point in time. This presumably explains why lottery winners don't usually experience a long-term change in their happiness rating.

So while genes give us set-points for levels of contentment within which we work over our lives – rather like a thermostat – at any given point in time non-genetic factors (i.e. family and social support) are just as important. However, contentment or happiness is just a figment of our imagination and personality. According to behavioural geneticists they are not specific or independent factors but just a composite of our personalities.[14] They can be predicted from studying the so-called 'Big Five' personality features: openness, conscientiousness, extroversion, agreeableness and neuroticism, which are themselves around 50 per cent genetic and 50 per cent environmental. Happy people generally have low scores for neuroticism, high scores for extroversion and conscientiousness, with increased openness and agreeableness scores helping a bit. Although happiness is elusive, it is still easier to ask how someone is feeling than to enquire about their neuroticism/agreeableness index. Luckily for authors of happiness books, the secret of what makes us regularly happy still remains unexplained.

Optimism is overrated

An important aspect of happiness is how we perceive the future – we differ in our optimism and pessimism. But, as with happiness, humans tend to overrate their own abilities: for example 85 per cent of people rate themselves as having above-average 'people skills' and 25 per cent rate themselves as exceptional – in the top 1 per cent. Over 90 per cent of us also believe we are above-average drivers, and these false assessments don't diminish with age.[15] Optimism and overrating our true abilities appears to be a common trait across cultures and part of being human. So is optimism a learned behaviour or genetic? A large Australian–Swedish twin study found (like previous studies)[16] a 40 per cent heritability for optimism in females with only a small – 10 per cent – effect in males, suggesting that men may be more easily influenced by outside factors.

Is it always good to be optimistic? Over 83 studies have been performed exploring how optimism influences health.[17] Most studies show that there is generally a slight (20 per cent) advantage to being optimistic in life when dealing with the after-effects of major life events or illnesses, such as bereavement, cancer or heart attacks.[18] Pessimism has, as expected, the opposite effect. But sometimes you can be too optimistic. Extreme optimists who guessed they would live 20 years more than national averages were compared with moderate optimists (the majority) who overestimated by a few years and pessimists who thought they would die early. The study found that the moderate optimists worked hard, saved well and smoked less, compared with the extreme overconfident optimists who worked less, saved less and smoked more.[19]

One of the longest longitudinal social studies, which are often more reliable, is called the Terman Study, named after the child psychologist who selected and tested 1,500 gifted high-IQ ten-year-old Californian children in 1921. Terman and his successors followed them all their lives, at first to see whether these high IQ kids would become remarkable adults, and later to record when and how they died.[20]

Of all the traits, conscientiousness best predicted longevity. But what was interesting was that the extreme optimists, particularly the men, died off earlier. Extreme optimists tend to be overconfident: because they never contemplate the consequences of failure, they are likely to take more risks, such as driving too fast, not getting medical check-ups, smoking, and not bothering to take medications. So could having a less rosy view on life sometimes be helpful?

The Stockdale paradox

Vice Admiral James Bond Stockdale was no wimp. As an officer in the United States Navy, he was awarded 26 personal combat medals, including the Medal of Honor and four Silver Stars. This made him one of the most highly decorated officers in US history and the highest-ranking naval officer held as a prisoner of the Vietnam War. While on a mission over North Vietnam in September 1965, he was taken captive and held as a POW. He remained in the Hoa Lo prison for seven years. There he was routinely tortured and beaten and mostly kept in solitary confinement. Sleep was difficult. His cell measured 3 square metres. A light bulb burned above him 24 hours a day and he was locked for 12 hours a day in leg irons.

When informed by his captors that he was to be paraded in public for propaganda, he told them that he would not allow this and so cut his scalp with a razor to disfigure himself. When they covered his wounded head with a hat, he beat himself with a stool until his face was so bloody and swollen he was unrecognisable, and so not much use for propaganda.

Stockdale survived for seven years.

When asked what his coping strategy was Stockdale replied: 'I never lost faith in the end of the story, I never doubted not only that I would get out, but also that I would prevail in the end and turn the experience into the defining event of my life, which, in retrospect, I would not trade.'[21] When asked about fellow prisoners who didn't make it, Stockdale replied: 'Oh, that's easy, the optimists. Oh, they were the ones who said: "We're going to be out by Christmas." And Christmas would come, and Christmas would go. Then they'd say: "We're going to be out by Easter." And Easter would come, and Easter would go. And then Thanksgiving, and then it would be Christmas again. And they died of a broken heart.

'This is a very important lesson. You must never confuse faith that you will prevail in the end – which you can never afford to lose – with the discipline to confront the most brutal facts of your current reality whatever they might be.'

Stockdale was by no means a pessimist, or he would have given up, but he clearly had a more realistic mild optimism that helped him survive.

Although humans have a clear inbuilt optimism bias there is a wide range of states that, being partly inherited, can be either useful or fatal in some scenarios. As the American writer William Arthur Ward put it succinctly: 'The pessimist complains about the wind; the optimist expects it to change; the realist adjusts the sails.'

Depression – or just an optimism deficiency?

Peter and Nigel were 42-year-old identical twins who were about to celebrate their joint birthday together with their family. But Peter hadn't answered the phone for two days and his family were beginning to worry. Eventually Nigel was woken by a late-night phone call: 'I'm sorry, we've found him.' Peter had hanged himself in his bedroom. The twins' birthday was spent going over funeral arrangements and working out how to disguise the rope marks around his neck. Peter had committed suicide after struggling with depression and alcohol for several years following his divorce. Both brothers always knew they had a tendency to depression which came from their mother's (Irish) side of the family, where several uncles had met the same end.

'Why him and not me?' Nigel asked. 'I should have done more. Yes, he was gloomy but I thought – like before – he would get through it. Sometimes I'd ring and he'd sound like his old self. Other times he was just full of negative thoughts and low self-esteem.'

So with the odds against him, why had Nigel so far avoided the depression that plagued his brother? Both brothers had the same genes and same family background. Their environments were similar but, crucially, not equal. Both tended to be moody and sometimes melancholy and had married around ten years before, but Peter's marriage subsequently fell apart – something he never quite recovered from. Nigel has taken the death of his brother badly and one year on still feels guilty, but he has not yet become clinically depressed and feels cautiously confident about his own future.

Ireland has for several generations had one of the highest rates of depression and suicide in the world, but it is not yet clear if genes or the Irish environment (weather and alcohol) are to blame.[22] Although finding causal genes even with large studies has not been easy,[23] both depression and suicide have a clear genetic component equating to about 40 to 50 per cent heritability.[24] And yet over half of identical twins with a depressed sibling are on average free of depression, so environment triggers are evidently also crucial. In the case of Peter and Nigel a stressful divorce may have been the factor that tipped the balance. Both twins are likely to have inherited a suite of genes that very slightly altered their brain chemistry, making them more susceptible to a rapid loss of optimism under stress. Dual twin suicides are not that uncommon.

Several large studies starting in the 1990s of a candidate gene (one that scientists suspect of being involved) have shown promise. They found that people carrying two copies (one from each parent) of a gene variant controlling transport of the key brain neurotransmitter serotonin (also known as 5-HTT) have twice the risk of major depressive episodes. However, as my colleagues at the Institute of Psychiatry found out, this seems only to hold true if they also suffer a major life crisis.[25] Since then, more than 55 other studies have attempted to copy the result and most (but not all) have succeeded – some in monkeys and rodents.[26] Others have used brain imaging to show that the anxiety and fear centre (the amygdala) lights up differently in those with this gene variant when stressed. Several studies implicated major stressful events, particularly childhood maltreatment and the response to serious medical illnesses.[27]

So if we are lucky we will have the right mix of genes so that our natural human optimism prevails and we will survive the many knocks and setbacks of life – illness, bereavement, disease. However, with the wrong combinations of genes plus the wrong circumstances, the natural protection is lost and the brain chemicals acting on our anxiety (amygdala) and emotional centres (rostral anterior cingulate gyrus) conspire to project a pessimistic, or some say a more realistic, view of the future.

However, there is a more intriguing possibility, which is that epigenetics is likely to play a role in this process. Evidence shows that life stresses may influence us by acting epigenetically on the genes. A study of young macaque monkeys showed that in those that had the key 5-HTT gene variant it was methylated and so inactivated when they were stressed. Monkeys with a different genetic make-up have a different stress response.[28] Other studies suggest that in humans, having a version of the gene alters your tendency to see the world through rose-tinted spectacles. The protective form of the gene lets you filter out or ignore negative images or thoughts, so protecting you from bad news.[29] This presents the possibility that in the future, when faced with a messy divorce, rather than resort to Prozac or gin and tonics we could take specific highly tailored chemicals to epigenetically steady our sensitive genes until the crisis has passed.

Genetic mindsets

Even environmentalist sceptics agree nowadays that many traits are to some extent heritable; out of hundreds of traits, we have found only a handful that are not. However we humans just don't like to be told that anything about us is predetermined. There is now some research evidence to suggest that telling people that their lives are in some way pre-ordained may actually be harmful and limiting.

The New York psychologist Carol Dweck has been looking at the powerful effect of mindsets. She took a group of ten-year-old schoolkids and tested them individually with the same simple puzzle. She then divided the class randomly, regardless of their real scores, into two groups and told individual students from one group that they had done well 'because they clearly had talent'. The other group were told they had done well 'because they had worked hard'. The groups were then asked if they wanted to take a harder test or retake the same test. The 'talented' group preferred to retake the same test; most of the 'hard workers' accepted the challenge to do a harder one. When both groups were tested on the new harder puzzle, the 'hard workers' beat the 'talented' group, and yet there was no academic or IQ difference between them – other than their mindset.[30]

This study showed the power of positive labelling to actually do harm: the talented group suddenly felt they were now failures. Negative labelling can be even worse. Tests have shown that even completing a simple check box at the top of a test indicating race or sex can lower subsequent test scores.[31] Not all of us are affected equally by these labels. Dweck has divided people into two broad mindset categories on the basis of answering a few simple questions. You can test your own mindset by looking at the extent to which you agree or disagree with these statements:

1. Your intelligence is something very basic about you that you can't change very much.
2. You can learn new things, but you can't really change how intelligent you are.

If you mostly agree rather than disagree with both statements then you are more likely to have a 'fixed' mindset. If you mostly disagree, you are more likely to have a 'growth' mindset. According to Dweck, fixed mindsets react badly to positive or negative labelling and have a fear of failure, whereas growth mindsets have a more pragmatic approach and will continue to take risks to expand their potential.

In the US in the 1960s kids were frequently lined up in class by their IQ scores, often with long-lasting adverse effects on both ends of the spectrum. As we enter the modern world of gene testing and potential genetic determinism, are we going to repeat the mistakes of the IQ test? While our personality genes make us more or less susceptible to having a fixed mindset or to being over-optimistic, this doesn't mean they can't change.

Most of us will perform better in exams if confidently told after a poor performance: 'You can definitely change and improve next time', rather than: 'Try to do your best, but don't worry, as you may have already reached your genetic potential.' So even if in the future your genetic make-up would be a rough guide to future performance, it would be counterproductive in terms of personal development to label anyone genetically talented. Positive Psychotherapy Intervention (using bio-feedback) pioneered by Marty Seligman has been used to alter mindsets in some people. Pessimists can become slightly more optimistic and improve their resilience and outlook on life, as well as reducing depression and improving their ability to withstand or recover from disease.[32] Although the approach has its sceptics, the US army has invested heavily in it and in 2009 gave $31 million to Seligman's team for their CSF (comprehensive soldier fitness) training programme to help fight depression, and improve mental toughness during trauma. Other cheaper self-help cognitive therapies are also increasingly popular. One example is www.moodscope.com, which is a free service that allows you to track your daily mood and assess your ups and downs.

An ancient approach to changing our emotional responses is meditation. Elite meditating Buddhist monks, after performing thousands of hours of meditation, can greatly alter their gamma brain activity.[33] Brain scans show amateur meditators can also structurally alter some brain areas like the hippocampus, affected in depression, after only eight weeks of meditating for 30 minutes per day.[34] All this points to the amazing flexibility, adaptability and neuro-plasticity of the human brain, which, with our rigid mindsets, we often ignore.

Another practical way to make yourself happier is to avoid the Victor Meldrews and move next door to happy neighbours. James Fowler studied 5,000 New Englanders over 20 years and plotted their changes in happiness against their social networks.[35] Happy and miserable people clustered together in ways that couldn't be explained by genes and families. This isn't just happy people avoiding grumpy people: the study suggested that happiness spreads like a virus. If you have just a single friend living within a mile who was sad and became happy, your chances of happiness increased by a quarter, and by more if they moved even closer. The effect was greatest for same-sex friends and extended to friends of friends of friends – the so-called three degrees of separation. The effect was most profound when you were at the epicentre of a happiness cluster and if your next-door neighbour was happy. Perhaps estate agents should now start providing neighbourhood happiness scores to raise property prices.

So while it is clear that genes are important for many facets of our lives, including happiness, all the evidence suggests that they no longer look like the dominant factor. We all have inherited sets of genes regulating optimism and pessimism, and they vary between us. However, we now know that both genes and their related mindsets, which we thought to be hard-wired, can be modified and reset along with the traits and personalities that define us as individuals.

In the next chapters we will learn more details of how and when these epigenetic modifications occur. As we will see, this also raises the possibility that you can pass on your acquired optimism or happiness to your children. But can you also pick up a new skill or talent and pass that to your kids?

3

'THE TALENT GENE'

Genius, motivation and taxi drivers

Being good-looking, fresh-faced and petite, with blue eyes and natural blonde cheerleader looks, it was never going to take long for identical twins Bryony and Kathryn Frost from the Isle of Wight to get noticed. Aged 18, their mirror-image good looks were set to make them millions in sponsorship deals and the tabloids loved them.

However, the Frost twins – or 'Frosties' as they are known – were also tipped as possible Olympic medallists in the gruelling steeplechase event. Their agent believed he had the equivalent of double Anna Kournikovas and the sponsorship offers started flooding in. Kathryn and Bryony were ranked first and second respectively in the UK Under 20s team and fifth and sixth in the world, and held the British junior 3,000m steeplechase record. 'We are just two young girls who love running and our dream is to go to the Olympics and win gold.' They started running and winning at the age of three in the school sports day and joined an athletics club at the age of ten. By 15 they had won three gold medals between them at the Island games in Sweden. They were dedicated and trained for hours every day, eating only fresh fruit and vegetables with the occasional small bit of steak to keep their weight down and improve their performance.

Sadly the twins never made it to Beijing. They had dropped their weight so much that they fell ill and found it impossible to regain their strength. They were soon diagnosed with anorexia nervosa – a common problem in female long-distance runners that also leads to hormonal problems, brittle bones and risk of fracture. They have now given up competitive running and their chances of Olympic golds. Using their unspent Adidas sponsorship money, they have started a property company together and hope to get rich instead.

The story of such similar talent in identical twins usually points us towards a major genetic influence. They had the same competitiveness, determination, small frame, physique and endurance that made them perfect middle-distance runners. But is most sporting talent genetic? A casual search on the Internet would suggest this is indeed the case, with a number of websites offering commercial genetic testing to assess your own or your gifted offspring's potential sporting prowess.[1] Companies include Atlas Sports Genetics,[2] who say that their $169 first sport gene test is 'geared specifically to show athletes, trainers and interested individuals where their genetic advantage lies'. So where does this evidence come from?

In 1998 a short article in the prestigious journal *Nature*[3] led to the media announcing that 'the gene for sports endurance' had been found. The team suggested that a gene called ACE, which alters dilation of blood vessels and blood pressure, can also affect endurance. Most genes have strange long names and scientists shorten them to make them more memorable and exciting. ACE is short for the Acetyl CholinEsterase gene, which produces a protein of the same name. They found that the gene variant ACE II was twice as common in 25 elite British mountaineers who regularly climb above 7,000 metres without oxygen as in nearly 2,000 normal men. They also found similar results in British army recruits who could do the most reps with dumbbells during the ten-week training course, and in another group with the best improvements in heart muscle.[4]

Soon after, another group claimed to have found 'the running gene'. Called ACTN3 (short for Alpha Actinin-3), this gene could alter endurance of muscle fibres by producing an enzyme of the same name.[5] We have two main kinds of muscle fibres: fast twitch fibres that don't require much oxygen and are useful for sprinting or power, and slow twitch muscle fibres for endurance running and sports for which oxygen is essential. Endurance runners have twice the frequency of this genetic variant as sprinters. They also found that mutant mice that lacked the 'endurance gene' caused their fast fibres to act like slow fibres and increased their treadmill endurance by 33 per cent. Human populations had walked (and sometimes run) out of East Africa 100,000 years ago following long droughts. In the exodus we diversified as we spread out in different environments. Because the gene frequency varied five- to tenfold across the world, the scientists behind the study hypothesised that evolution could explain these differences in sporty genes.[6]

The so-called running and endurance genes caught the prevailing mood – it suddenly seemed obvious why athletes of African origin had won every track gold medal in long distance and sprinting since the 1970s. All the best sprinters originally came from West Africa, and long-distance runners from East Africa. If we could identify the genes of the large numbers of Kenyan medallists, who all lived in a tiny area in the Rift Valley, we thought we would have the key to sporting success. Around the world sports academies started looking at genetic testing of their most promising youngsters, while others started touring East Africa with syringes and test-tubes.

But while millions of dollars were invested in biotech companies and sports genetics grants, a slight problem emerged. The data of follow-up studies was not quite so solid. After the initial positive research papers featured by the media, more studies were performed, with better designs and much bigger numbers, which produced negative results. By 2008 over 50 studies had been carried out with athletes and these two genes, and it became clear that (like many other hyped candidate-gene studies) there are no consistent strong effects.[7] With international consortia we have recently also failed to find any effect of these genes on muscle bulk or muscle strength using tens of thousands of subjects. So, although we cannot rule out tiny effects, for prediction of athletic ability they are pretty much useless – and certainly can't beat a stopwatch.

Tiger mothers – and Tiger Woods

Vanessa Mae, the child-prodigy violinist turned pop star, began playing piano at the age of three and violin at the age of five. According to the *Guinness Book of World Records*, at 13 she was the youngest soloist to record both the Beethoven and Tchaikovsky violin concertos. At ten, she made her international professional debut at the Schleswig-Holstein Music Festival in Germany and with the Philharmonic Orchestra in London. On entering adolescence Vanessa Mae broke away from her traditional classical influences and became known for her sexy music videos, playing electric violin and promoting her best-selling albums. She performed in the interval of the Eurovision song contest watched by billions. In 2006 she was ranked as the wealthiest young entertainer under 30 in the UK, with an estimated fortune of about £32 million stemming from concerts and record sales of over 10 million copies worldwide. Where did this amazing talent and success predominantly come from? Her genes or her training?[8]

A few years ago I was interviewed with Vanessa Mae, then aged 31, for a TV programme about her life and the role of nature and nurture in her success. She did not like my opinion at the time that good genes were the prevailing reason she was so talented. She was adamant that her success was due to years of dedication and hard work from an early age.

She was brought up by a strict Chinese mum, who, in the style popularised by Amy Chua's *Battle Hymn of the Tiger Mother*, early on enforced a strict regimen of social isolation and practice on Vanessa. The practice paid off – but at a cost. Vanessa was an only child with an absent father, and was not allowed to go to a normal school and have regular friends. Her mother, herself an accomplished concert pianist, was such an obsessional disciplined trainer that it caused major scars in their relationship and they no longer speak. Her experiences, isolated childhood and adolescence have clearly left a mark on her.

A recent wave of excellent best-selling books – Malcolm Gladwell's *Outliers*, Matthew Syed's *Bounce*, Geoff Colvin's *Talent Is Overrated*, David Shenk's *The Genius in All of Us*, Daniel Coyle's *The Talent Code* – have all provided the same clear message: that there is actually no such thing as innate talent.[9] They all quote the work of Anders Ericcson, a Swedish psychologist working in Florida, who did much of the key research.[10] He is known as an expertise-ologist. The essence of his work is that our society vastly overestimates and applauds talent and ignores hard work and graft. He used both experiments and historical examples to make his point.

Wolfgang Amadeus Mozart is to most people the clear example of a child genius. By age six he had started to compose piano and violin pieces and was performing regularly. But what is often forgotten is that his father, Leopold, was one of the best music teachers of his time and had started training young Wolfgang intensively from age three. It probably wasn't until age 21, after 18 years of gruelling practice, 12 hours a day, that he produced his finest work – his violin concerto No. 9. When looked at objectively by experts his early child works, written between 11 and 15, show no particular originality or genius; they are merely good rearrangements of other composers.

Ericsson's view is that only hours of gruelling practice make someone good into someone exceptional.[11] Mozart is estimated to have put in around 3,500 hours of practice before he was six and over 10,000 hours by age ten. He himself said: 'Nobody has devoted so much time and thought to composition as I.'[12] Using more modern musical examples (as highlighted by Gladwell), The Beatles were far from overnight sensations: they spent several years playing together for 12 hours a day in a cellar in Hamburg to hone their skills and reach the magic threshold of over 10,000 hours of practice.

What about sport? Surely that is where natural talent is most obvious. Tiger Woods is hailed as the most fluid 'natural' golfer of his generation. Yet, like Mozart, he had a great coach, a dad who was also an ex-sportsman. Tiger started playing golf at the age of 18 months and entered his first tournament (a pitch and putt competition) at two. He was clocking up incredible hours of practice – over 10,000 – before he reached 16, and six years before he became the youngest winner of the US Masters aged only 22. The years of isolation and practice may have made him a great golfer, and the richest sportsman in the world, reputedly earning $90 million each year. Money however may not have made him happy or a rounded individual, as seen by his strange obsession with infidelity, hookers and Las Vegas.

If talents like Tiger Woods are so rare, these genetic miracles should only come along once every few generations. But an unassuming 21-year-old Northern Irishman, Rory McIlroy, recently beat even his amazing record. McIlroy easily won the US Open tournament in 2011 after the traumatic experience of having 'choked' as his nerves got to him while leading on the final round of the US Masters at Augusta a few weeks earlier. He attributed his ultimate success to a new mental attitude: putting golf into perspective. He spent three days in Haiti before the tournament, where on that hurricane-devastated island 50,000 people were now living in tents on a nine-hole golf course. He was the youngest winner for nearly a century and had a record-breaking final score. Just four years after turning professional he was ranked number one in the world. Surely his gift and rapid overnight success was God- or gene-given talent?

But Rory was not a one-off. He came from a golfing family and his father was an excellent scratch golfer. Like Tiger he was an only child and lived his early life at the local golf club near Holywood outside Belfast, where he was the youngest full member aged seven, but still wasn't allowed in the bar. His father started practising with him from 18 months, and by two years old he could drive the ball over 40 yards. He won his first international tournament at age ten. He slept with his favourite toy – his driver. By the age of 18, when he turned professional, Rory too had clocked up way over 10,000 hours of gruelling practice – probably more important than small variations in his 25,000 genes.

The myth of the power of genes for elite runners has also been challenged.[13] Ninety per cent of elite Kenyan runners come from the same tiny area of the Rift Valley near a small town called Eldoret and belong to the same tribe, the Kalenjin. However, unexpectedly, they were actually not generally related to each other, but did have a few unusual environmental factors in common. They lived at altitude all their lives, which increased the number of red blood cells circulating oxygen naturally. They also ran to school every day in their school uniforms – an average of eight to ten miles a day. So again by the age of 18 they had accumulated vast hours of running, which felt natural for them. At the time a US car bumper sticker read 'Give our athletes a chance – donate school buses to Kenya'.

Haile Gebrselassie, the world record holder in the marathon and perhaps the greatest distance runner ever, was not Kenyan – he was Ethiopian. Although he too ran to school from the age of five, despite his skin colour his genes, like most of his countrymen's, are much more similar to Europeans' than to Kenyans'. While we are readily biased by the colour of someone's skin when predicting their physical or intellectual abilities, surprisingly skin colour is controlled by just a handful of genes, and is a poor guide to the other 25,000 underneath. Indeed there is more genetic diversity in one small area of Africa than there is in the whole of Europe.

The idea that humans have really diversified genetically in our running skills has also now been questioned. One current theory of our evolution and separation from other primates 6 million years ago is that all humans needed to run for long distances on two legs. This was both to escape predators and to catch prey on wide hot plains. Although compared with most animals we are pretty useless sprinters, over 20 miles or so in the heat we are nearly unbeatable. In some regular long-distance contests, men can – and do – beat horses. This suggests that it is a hard-wired human trait, not a rare gift of a few.[14]

The Knowledge: talent or experience?

The expertise camp give many other examples of hard work trumping talent in other areas: in art the late-stage improvements to the work of Picasso and Cézanne; in literature the case of Mark Twain, who took ten years to finish *Huckleberry Finn* – here the great 'talent' shone through only after they had been practising for a very long time. In contrast there are very few examples of success due to innate gifts and little practice – other than perhaps underwear modelling, or being on *The X-Factor*.

Black cab taxi drivers in London are unique in a number of ways, one of which is that as adults they have to pass an intensive memory test. They have to spend an average of two to four years travelling around on scooters visually memorising in their heads 25,000 routes and alternatives from one place to another across large distances without GPS. This intensive test is called 'The Knowledge', and many fail the exam even after ten years of trying. So were the successful cabbies more talented? Researchers at the Institute of Neurology in London scanned their brains and found that the area underneath the cerebral lobes called the posterior hippocampus actually had 50 per cent more grey matter than in bus drivers or normal controls. In contrast the part in front, called the anterior hippocampus, was smaller.[15]

So did they have a naturally large hippocampus – the brain area responsible for spatial imaging and memory?

When they retested retired cabbies a few years later they found that the hippocampus had actually reduced in size, showing that it was the hours of daily practice, not genes or IQ, that had made bits of their brain reversibly larger.[16] This extra growth of the brain was akin to a body-builder increasing their biceps, and allowed them to memorise thousands of alternate routes, while generously giving their strong opinions to passengers on any subject. The same is unlikely to be true of most untrained New York drivers, who often get lost. Expertise in other domains such as music, mathematics and linguistics has also accompanied relative increases in grey matter in some parts of the brain. However, as with most things in life, there may be a price to pay. Studies also showed that as some areas improve, some other bits of the brain suffer: cab drivers, for example, were less good at some other memory tasks, perhaps explaining how they can forget to give change.

What about other cerebral professions? Doctors, who have to learn lots of facts over many years and have above-average IQs, did not (despite their big heads) have any obvious brain-size differences compared with controls. What about in more real-life scenarios? Is the talented youngster sometimes going to outperform the older experienced doctor? Anders Ericsson and his fellow psychologist Paul Ward run courses in Florida where they try out experts and novices in pressure situations – for instance in a hospital nursing scenario, trying to save a dummy patient from dying. 'That's when we uncover the expert superiority: their ability to perceive more information, and also, after the fact, remember more of the thought processes than the novices,' says Ericsson. 'Some key differences would be the way in which they pick up information from the environment,' Ward says, 'and the way in which they comprehend that information such that they could then use it to good effect.'

Ericsson and Ward have used techniques like this to compare thousands of experts with novices in fields from music and sports to medicine and law enforcement. So far they've found no evidence that experts are born with any more natural 'talent' than other people. 'We have yet to find any compelling evidence that any talent matters,' says Ericsson. 'Anyone with the right kind of practice will be able to dramatically improve their performance and it looks like they would be able to become experts with sufficient practice.' The key they say is focusing on the activity you are weak at. Ericsson's team famously trained a randomly selected 'average' student called SF to become a memory expert. They did this in just 250 hours of training in the lab for an average of one hour a day over two years.[17] Most people are hard-pushed to remember phone numbers of over eight digits. By the end he could memorise over 80 digits correctly. If Ericsson and Ward are right, any dedicated parent would, if they started early enough, be able to produce a prodigy – even if only in reading the phone book.

In 1967, a young and newly married Hungarian educational psychologist called Laszlo declared that he was going to produce children who would be chess champions.[18] He chose this he said because it was a sport or skill that was easy to grade. Two years later his first daughter Susan was born. He started playing chess with her when she was three, using fun techniques to get her hooked. It worked, and she had accumulated hundreds of hours of practice before the age of five. Two more daughters followed and the same regimen was applied. All had reached the magic 10,000 hours by the age of 14, and all became chess champions. One was the youngest grandmaster ever; another won eight games in a row against male grandmasters; and the third became the only female winner of the under-twelve world championships. Some tough critics argued that as he was a reasonable chess player himself, this could be partly genetic. So in response he enthusiastically offered to adopt three new children and repeat the process. Sadly for him and for us, his wife felt a bit tired of chess and children and overruled him.

But can this really be true? Can any child be trained to be the best chess player, violinist, tennis player, golfer or mathematician? Are we returning to the blank slate idea and the Jesuit mantra? (St Ignatius Loyola founded the Jesuit order, with its famous maxim: 'Give me the child until he is seven, and I will show you the man.') Since the Second World War, apart from a few blips, the nurture 'blank slate' concept has been dominant. The idea is usually traced to the seventeenth-century philosopher (and parental expert) John Locke, who said: 'Let's suppose the mind to be, as we say, white paper void of all characters, without any ideas. How comes it to be furnished? . . . To this I answer in one word, from EXPERIENCE.'

The idea fitted the prevailing mood of the Enlightenment, challenging the supposedly innate authority of monarchy and class and allowing expression of personal freedom. An influential Victorian with the opposite view about nature and nurture was Darwin's half-cousin, Francis.

The ultimate Victorian gentleman scientist and explorer, Francis Galton is credited with being the father of twin studies and heredity (as well as discovering African countries, fingerprints, regression statistics, and anticyclones on the way). He took a particular interest in geniuses, and in 1869 he published a best-selling book, *Hereditary Genius*, based on his research on the 400 most eminent men of Britain in the 1860s.[19] When he studied the top 'eminences', using his personal and debatable choice of graduates of Cambridge and Oxford as well as judges, he found that they were related more than you would expect by just chance alone. He concluded that his results showed: 'It would be quite practicable to produce a highly gifted race of men by judicious marriages over consecutive generations. Each generation has enormous power over the natural gifts of those that follow and maintain that it is a duty we owe to humanity to investigate the range of that power and exercise it in a way that without being unwise towards ourselves, shall be the most advantageous to future inhabitants of the earth.'

The subsequent enthusiasts of the eugenics movement of the 1930s suggested theoretically that selectively breeding champions would produce a super-race. However, human experiments to actually produce 'master-races', such as the Nazi Lebensborn project or the early Aryan Kultur settlements in Nueva Germania, Paraguay, in 1887, ended as dismal failures. Even if we ignore the practicalities, is there a basis for Galton's view of hereditary genius? As we have discussed before, the results of the twin and adoption studies of IQ are very consistent, showing an average heritability of around 60 per cent in over 30,000 individuals. So clearly genes must have some role, in IQ at least.

Does practice makes perfect pitch?

What do genetic studies in the population tell us about sporting abilities? A massive study of 37,000 pairs of European twins set up by the sadly deceased Finnish geneticist Leena Peltonen showed clear results. In all of our seven twin registers tested, participation in sports was influenced 70 per cent by your genes after the age of 21.[20] Before then, school and parents controlled your actions to a large extent, but as they left home people returned to their genetic tendencies of either natural laziness or sportiness.

So if willingness to play sports or exercise regularly is quite strongly genetic,[21] what about sporting ability? With Dutch colleagues we looked at our twin registries a few years ago. In over 4,500 twins we identified sporting skill in over three hundred who had competed for their county or nationally at any of 20 sports. We found a 66 per cent genetic influence on competitive ability at any sport.[22] However, they often didn't share the same sporting prowess and, for example, only 50 per cent of identical twins were both competitive at tennis. The heritable results for ability were similar to our estimates for the structural elements: lung capacity, muscle strength and muscle bulk.[23] More recently with Cambridge scientists in the Actiheart study we have found that cardiac fitness is also heritable.

The work of Eriksson and others focusing on expert training and long-term practice ignores the fact that not all people given the same training regimen will respond in the same positive way or as fast. Some are more likely to then give up disheartened, while others will see quick results and keep going. This is often the problem of taking anecdotal sporting examples. You only see the successes. Who knows how many dads gave their kids a golf club driver for their second birthday who never became a Tiger Woods or a Rory McIlroy? Ninety non-sporty overweight couch-potato Louisiana families in the Heritage study were put through a 20-week training programme using indoor cycling machines. As you can imagine there was a wide variety in fitness response. There was a clear heritable component (50 per cent) in the rate at which their oxygen uptake improved. This was regardless of how fat or fit they were when they started.

Being a talented musician is a more complex skill than cycling or running, involving a number of different skills in addition to manual dexterity and coordination. In trying to separate the key components of what makes a talented musician we performed some twin studies with the help of a gifted NIH researcher, Dennis Drayna, who had designed a cunning test of relative pitch perception. It is called the distorted tunes test and comprises 24 well-known tunes like 'Yankee Doodle Dandy', 'Fleur de Lys' and 'London's Burning'.[24] Roughly half have deliberate subtle and obvious errors introduced into them, and the test subjects have to guess which are correct or not. We tested our twins and found that around 10 per cent had perfect scores, most scored about 85–90 per cent correct and 8 per cent hadn't a clue, scoring no better than chance.

This tone-deaf group are responsible for much global pain and suffering in karaoke bars. Clearly if you were tone-deaf it would be nigh impossible to be a concert pianist, and very difficult if you were not perfect.[25] But can this skill be acquired in all kids by intensive early training, as in the chess family? This seems unlikely in tone-deaf kids, as the trait was very similar in identical twins with a high heritable factor of over 80 per cent. We also found several pairs of twins who both had top scores but had never played an instrument in their lives, as their parents were either poor or uninterested in music. One recent and as yet unreplicated study of Finish families has suggested that a propensity to listen to music as well as creativity was heritable, and much more speculatively that genes for sociability (vasopressin) might be involved.[26]

Absolute pitch (AP), also known also as perfect pitch,[27] is the much rarer ability (found in around 1 in 1,000 to 10,000 people) to name the pitch class of any given tone from any instrument or sound without a reference).[28] This ability is thought to be a 'musical gift' partly because of its possession by musical talents such as Mozart or Stevie Wonder.[29] Increased frequencies of AP of up to 10 per cent have been documented among classes of gifted young musicians, showing that it is possibly helpful, but certainly not essential. Mild autistics also have higher rates of AP, and brain scans have shown that those with AP have increased brain cortical connectivity.

There have been many arguments over the role and timing of early musical training in acquiring AP. A recent small twin study of absolute pitch found a 79 per cent agreement in the 14 monozygotic pairs and a rough heritability estimate of 66 per cent, suggesting some genetic influence is present. But this clearly doesn't rule out the environment. AP is also more common in countries with tonal languages (China and Vietnam) and decreases when these populations emigrate to the US or Europe – again suggesting a key influence of early sound exposure. There are even training schools like that of David Burge and www.perfectpitch.com, which will train you to attain AP at home for a mere $169 fee. Sadly, while there may be improvements in naming notes and tones, there are few if any examples of students that have mastered AP to a high level this way. However if you have the cash it might still be a useful trick at parties.

Boxing, belief and willpower

Despite a great deal of conflicting information presented by proponents of both the training and the genetic camps, both groups tend to ignore the existence of the other. Until now we have talked mostly about skills and expertise in muscles, coordination and brain-power and the ability to practise from an early age. While there is undoubtedly some genetic component to talent, exactly what the key set of genes do is far from clear. Genes controlling muscles, neurones, eyesight, hearing or coordination on their own are not enough. There are clearly other factors that differentiate the great from the good. The most famous ice hockey player of all time, Wayne Gretzky, famously said: 'Maybe it wasn't talent the Lord gave me, maybe it was the passion.' What form could this passion take?

The ability to have a strong faith can be beneficial for survival. Many sportsmen have this same conviction and faith in their own abilities that helped them with success, even when the facts pointed elsewhere. Jonathan Edwards, the gold medal-winning triple jumper and religious TV show presenter, suddenly and publicly gave up being a devout Christian on his retirement from athletics. Perhaps he no longer needed the same conviction or divine help to drive him. Others like Matthew Syed, the champion table-tennis player, have pointed out that successful people seek winning the way a drug addict seeks heroin. The effect of the win, like that of some drugs, is so short-lived that as soon as they have left the winner's podium they feel depressed and unfocused and need to start over again immediately. Other differences may be more subtle.

Born 20 minutes apart in Lewisham, a poor part of southeast London, on 3 May 1934 were two identical twin boys. Their father was a keen boxing fan, and aged 14 they joined the local Bellingham amateur boxing club together. They were strikingly similar, with strong foreheads and deep-set eyes. They were tall (both 6′ 2″) and powerful but nimble heavyweights. They were inseparable and George would always fight second so that his twin wouldn't get upset watching him. Though gifted, they found their way slowly, both losing their first four amateur fights, until Henry started winning and got selected for the Olympic team. Both were given professional contracts in a blaze of publicity in 1954, when they sparred together in front of TV cameras.

Henry kept winning British, European and Commonwealth titles, and his career reached a historical moment at Wembley Stadium in 1963 when in round 4, with a vicious left uppercut, he knocked out Cassius Clay (later to become Mohammed Ali). Sadly for him, Ali was saved by the bell, and dodgy delaying tactics used by his trainer Angelo Dundee. Ali recovered and came back at Henry with a flurry of blows, opening up nasty cuts to his head which stopped the fight. Ali said later: 'He hit me so hard – even my ancestors in Africa felt it.'

So while Henry is remembered as one of the best boxers Britain has ever produced, what happened to twin brother George? Although no slouch, he never found the same level of success. As a pro he won 16 fights and lost 14 and then retired. One theory for his relative lack of success was his (genetic) susceptibility to facial cuts, but probably he was getting hit more than his brother. As in many other cases, it's hard to find much difference between the twins to help explain this subtle but vital ingredient of being just good and being at the top of the game.

However, there are always random events that we can implicate. For instance an attack of rheumatic fever at age 16 stopped George training, perhaps at a critical point. Or again we might point to the skill (or lack of it) in a junior doctor making some extra cash, who helped stitch up cuts on both the twins at the ringside. I say this because my father claimed to have done this, and he, like me, through too little practice – or talent – was a terrible surgeon.

However a more likely explanation to me is a very slight difference in motivation between the Coopers. Ultimately Henry probably just had the extra determination and wanted to win a tiny bit more. Considering that there are millions of identical twin pairs in the world, there are relatively few of whom both have reached or had the motivation to reach the top in similar high-profile areas. Examples include the non-famous or more secretive twin siblings of Mario Andretti, Jerry Hall, Keith Chegwin, Joseph Fiennes, Isabella Rossellini or Curtis Strange. We will return to the question of why determination may differ later.

Standardised IQ tests for intelligence have been around for nearly a century since Alfred Binet in France and Lewis Terman in the US started using them as predictors of talent. Although twin and adoption studies have consistently shown a clear but politically controversial genetic component, a number of odd findings have emerged which question IQ's value. One is the fact that every decade in every country tested, the average test results improve and continue to do so. Another is the disappointing result that Terman had when he followed his most gifted children in California. Very few of them ended up the successes he would have predicted from their high IQ. More puzzling is a study that showed that the strong genetic IQ effect disappeared towards zero in very poor and deprived children.[30]

The important ingredient could be motivation. IQ tests were recently looked at in 2,000 kids and researchers found that scoring well in the test was strongly related to levels of personal motivation.[31] Those who scored worst also had relatively lower motivation levels, so failure became a self-fulfilling prophecy. Those who do well may just be the ones trying to impress the most. IQ tests (and according to some, life itself) can be seen as a test of both IQ and personality, and possibly also circumstance and environment.

Motivation is thus a key underestimated factor in success. It is likely to be crucial in being talented, making the difference between the child who can sustain the hours of tedious training and the one who gets distracted and loses heart. Could this, rather than strength, reflexes, eyesight or manual dexterity or perfect pitch be the missing factor?[32] Over 35 years ago a near-forgotten study looked at 61 pairs of twin schoolgirls and found a clear genetic influence on motivation – so the genes for this trait could be the most important genes of all.[33]

The pro-training camp often forget that by only looking retrospectively at the successes you don't see how they have been slowly selected for this trait. Nor can you see all the others who gave up years before, demotivated. More often than not one of the parents had the same steely determination, even if they never practised the same skill. So the key motivation factor is again likely to be a mix of genes and environment.

What happened to the offspring of the cohort of medal-winning Kenyan athletes of the 70s and 80s? Well, the trophy cupboards were bare in the 1990s. The genes on their own were not enough, and the next generation didn't produce any prodigies or win any medals at all. Perhaps because the medal-winning family had prospered, the drive and hunger to succeed was now gone. Or perhaps the fame and riches meant they no longer had to run to school every day?

Muscle memories

Where there is a clear interplay of genes and environment (in this case training and practice) there is an obvious role for epigenetics. Is there any evidence that extended practice can modify your genes? A recent (as yet unreplicated) Italian study showed that athletes were more likely to possess certain key gene variants that made their genomes less likely to methylate. This suggests it made their genes generally harder to switch off.[34] This lower level of methylation had the effect of stimulating muscle cells and muscle growth more than average in response to exercise. The relative importance of genes and training in this epigenetic process is still unclear, but early signs are emerging of potentially important different levels of gene methylation between physically active subjects and couch potatoes.[35]

Studies in rats have attempted to imitate the effect of giving out free gym memberships.[36] Into some lucky rats' cages they put a circular treadmill. Rats love to run. In fact they found they ran on average 10 miles every night just for fun. Previous studies had shown that rats and mice vary naturally in their normal running levels, depending on their genetic background and gender.[37] When after four weeks the now seasoned runners were compared against the non-gym rats they found they performed much better in tests of physique and response to stress.

That wasn't a particular shock, but what happened in the brain was. The genes in a key part of their brain, the dentate gyrus, had been modified epigenetically. This is a part of the brain in the cerebellum that controls muscle movement. There was a clear methylation difference between the rat groups. Studies are still ongoing, but it is likely that other genes are altered, and although there are practical problems in performing brain biopsies in humans using gyms, it is likely that similar changes occur. Some proportion of the changes to genes caused by exercising the muscles or the brain in adults could be retained as genetic marks, and passed on to our children to either use or waste.

It is hard to test muscle or activity differences across generations, but testing memories might be more feasible. As proteins only survive intact in our bodies for up to 24 hours and neurons can't divide and replicate, exactly how memories have been retained for years has been a mystery. Recent exciting research from a group in Alabama has found that when rats are given powerful memories – usually conditioned responses to a shock – these short-term memories are stored briefly in our old friend the hippocampus. They are then passed to the cerebral cortex, where they stay for years. This memory is regulated by long-lasting epigenetic mechanisms. Experiments show that these painful memories in rats can actually be erased by chemicals that block methylation (such as DNMT inhibitors). There is increasing evidence that musical training may have its long term effect on the brain and musical memories and skills via epigenetics.[38] Epigenetics is thus a great explanatory mechanism for both reversibly modifying our genes and producing (and allowing us to forget) our memories.[39]

What of designing future talent and elite status? One ingredient for success is the ability to pick star kids early, and gene testing was the early hope. We have seen how gene testing has been oversold and is still decades away – if achievable at all. But the following factors will certainly increase your chances. First, having a determined, highly motivated parent with whom you share the same character and similar genes. Second, start training early in life or live a long time. Third, acquire a strong faith in your own ability. Fourth, work on improving your willpower.[40] And finally get lucky, as most of the world's most talented people go unnoticed.

If you have all that burning ambition and all else fails, you could always try drugs. Doping is common in many sports and many believe that reversibly changing genes by the use of epigenetic drugs is likely to be very hard to detect, safer than steroids and possibly the sad future of many sports.[41] Cheating in exams or training with epigenetic smart drugs to improve memory is another option for the truly ambitious. The more serious upside is that epigenetic drugs also have great potential in treating dementia and Alzheimer's disease,[42] which one in five of us will suffer from.[43]

The fact is that despite all the best efforts, training, encouragement, similar genes and environment, offspring are more often than not a disappointment to their gifted genius parents, whether they are racehorses, scientists, musicians, writers or athletes. Nature just seems to have a way of balancing the books and keeping us guessing on what it produces. Einstein said in 1952: 'I have no special talents. I am only passionately curious.' In the next chapter we explore one of the great human passions: religion.

4

'THE GOD GENE'

Genesis, Jedi and Hollywood

Several million years of evolution have given us a tendency to feel content and be mildly optimistic for the future, but until the last 100 years our expected lifespan was around 50 years at best. So what kept our ancestors going over millennia when faced with so much disease and imminent death? I'm not sure that mild unfocused optimism alone would have been enough. Perhaps real faith was needed and the search for God and belief in the welcoming afterlife was the answer.

One of the surprising results of the 2001 UK census was that 390,000 citizens identified their religion as Jedi. This is a religion invented by George Lucas as a composite of Eastern cults and religions that worship the power of nature – 'The Force'. Although most respondents were undoubtedly joking, over 16,000 members have signed up to its serious side. There are eight Jedi chapters worldwide committed to its aims of purity and charity and respect for nature. It is led by its founder Daniel Jones, and after recognition by the UK census office, is currently the fourth-largest religious group in the UK and very likely to become third in the latest census. What does this tell us about the state of religion in the West? Is it changing? Or is it even disappearing? Have modern lifestyles changed the meaning of belonging to a religion, and is a belief in powers greater than ourselves hard-wired into most humans?

Identical twin sisters Debbie and Sharon were adopted and

separated soon after birth in New Jersey. They were raised apart in similar families, but very different religious environments – one in an orthodox Jewish family where the father was a psychologist, the other in a Catholic family where the father was a furniture salesman. Both went to regular services and Sunday school and were firm believers in God. They both independently went to college and both studied social work.

Debbie married a Jew and moved to Connecticut; Sharon married a Catholic and moved south to Kentucky; both had kids. Debbie eventually found out about her adoption and hired a private detective to track down her biological family, who told her about her twin. She discovered that her mother was originally German – probably half Protestant, half Jewish – and her father probably an Italian Catholic. When the twins were finally reunited in their forties, Debbie remembers: 'I was struck by the fact that we didn't look as alike as I thought we would. We had very different hairstyles and clothes – but we had many other odd habits in common, such as how we moved our hands and made faces. Everyone told us we both went a bit cross-eyed and rolled our eyes when we got excited.'

Soon after their reunion Sharon accompanied her sister to the conservative Jewish synagogue and said: 'It felt so natural to me,' and Debbie felt similarly relaxed in her sister's church and congregation. Both now help out with each other's charity work and remark on the similarities of their religious activities. Neither sister wanted to change her religion, or to convert the other, but their hairstyles and choice of clothes have slowly become more similar.

The only area they have difficulty discussing came up on a visit to a US theme park with dinosaurs. This was the tricky area of creationism and the interpretation of the book of Genesis. Debbie believed in Darwin's theory of evolution and Sharon did not. Sharon on moving south had abandoned Catholicism and had joined a fast-growing 'and more positive and practical' local Evangelical church. She now accepted the teachings of her church, which took literally the words of the book of Genesis, suggesting the earth was less than 8,000 years old. To Sharon it was quite possible that dinosaurs and man had coexisted at the same time, and she didn't like discussing areas that might contradict the teachings of her church, which she was proud to belong to. To Debbie, being Jewish, science and faith had a more flexible relationship which normally would be freely discussed. But she knew it made her twin uncomfortable.

As they were identical twins, only culture and family environment should in theory explain their religious differences and behaviours. Clearly both twins had a similar predisposition to believe in God and attend and participate in religious activities and charity work and derived great pleasure from it. Both were also, luckily, very tolerant. These traits could have been socially or genetically influenced, as they were raised in religious and happy caring families. Yet neither of them wanted to convert, and they clearly differed in their specific beliefs and affiliations – which must presumably be due to their environment.

Before we delve further into what makes some of us religious and others not, it is important to clarify what we mean by the loose term 'religiousness' or 'religiosity'. It has at least three dimensions, known as the 'Three Bs': belonging (affiliation), behaving (attendance), and believing.[1] Although they are somewhat linked, some believers are non-attenders and vice versa, and studies have also shown differences in behaviour between the groups. So exactly how you are religious matters as much as how religious you are. For the genetic studies we discuss, if you firmly believe in God your precise belief in religious stories or the nature of God is largely immaterial.

Despite the three dominant monotheistic religions emanating from the same roots of Judaism and Abraham around 4,000 years ago, humans are very diverse in our religious beliefs and make a big play to accentuate them. Over two-thirds of the world's population believe today that religion is important to them.[2] Faith is sometimes described as a strong belief in something that cannot be demonstrably proven. In the New Testament, faith is simply described as 'the substance of things hoped for, the evidence of things not seen'.[3] So some people's faith is to believe literally in the Quran being God's word (as do most Muslims); that the world is only 8,000 years old (as do 45 per cent of Americans); or that man is descended directly from Adam and Eve (60 per cent of Americans).[4] Others with a vaguer knowledge of history believe that Joan of Arc was Noah's wife (10 per cent of Americans),[5] or those with a longer cosmic perspective may believe more in a God that designed the universe, with its sextillion stars, well before the Big Bang 14 billion years ago. Whatever your faith – and we probably all have some – it is important to appreciate our wide cultural and educational differences and contemplate where they come from.

'The faith gene'

I am frequently asked by journalists to recall the most surprising finding of our twin studies. The study of religion and belief in God is the one that always comes to mind, and the results are not easily accepted by many people. Most people can accept diseases or height and even weight being genetically heritable to some extent, but when it comes to our personal beliefs we tend to be more sceptical. For many, the idea that there is a genetic component to our faith – or lack of it – is a stretch too far and damages the concept of self-determination that we hold so dear. Nevertheless science has shown us clearly that our level of belief in God and overall spirituality is shaped not only by a mix of family environment and upbringing – which is not surprising – but also by our genes. Twin studies conducted around the world in the US, the Netherlands and Australia as well as ours in the UK show a 40–50 per cent genetic component to belief in God.[6]

What is striking is that these findings of a genetic basis for belief are consistent even across countries like the US and the UK, with their huge differences in beliefs and church attendance. For example in the latest surveys in the US, when asked, 61 per cent of white Americans say they firmly (i.e. without any doubt) believe in God, compared with only 17 per cent of firm believers in similar populations in the UK – greater than a threefold difference. The opposite scenario of non-belief is also true – only a tiny 3 per cent of the US population report being firmly atheist compared with 18 per cent in the UK. As well as belief, participation follows separate trends in the two countries. Some form of weekly church attendance is now nearly three times higher in the US than the UK.[7]

Sceptics among you might say that the twin studies showing similarity for belief are just reflecting some cultural or family influence that wasn't properly corrected for in the study design. However in the Minnesota adopted-twin study mentioned previously they also looked at religious belief in a number of adopted twins raised apart like Sharon and Debbie. They found exactly the same result – greater similarity in identical twin pairs, even if raised apart. The conclusion is unavoidable: faith is definitely influenced by the genes.

To uncover in more detail exactly what part of belief or religion was genetic, an unlikely research partnership was formed between two academic twin experts – Nick Martin, an extrovert atheist Australian, and Lindon Eaves, a British lay preacher originally from Birmingham.

In an attempt to separate the '3 Bs' they asked a range of questions attempting to get a handle on individual differences in spirituality. They defined this as 'the capacity to reach out beyond oneself and discover or make meaning of experience through broadened perspectives and behaviour'. The scale is based on three main factors: self-forgetfulness, transpersonal identification and mysticism. Questions in the test they designed included:

'I believe that all life depends on some spiritual order or power that cannot be completely explained' – true or false?

'Often when I look at an ordinary thing, something wonderful happens – I get the feeling that I am seeing it fresh for the first time' – true or false?

They estimated the heritability of 'spirituality' to be around 40–50 per cent, which is quite high considering how tricky it is to measure.[8] Other US studies using even more detailed questions in larger numbers have found similar or even stronger genetic influences.[9] These studies demonstrate our variable but innate inherited sense of spirituality, which affects how we perceive the world, ourselves and the universe. This is independent of our formal religious beliefs and practices and, strangely, largely independent of family influence.

The positive feedback and inner reward we get from these spiritual or religious thoughts could also account for some differences. One individual during prayer or meditation may feel a rush of immense joy and fulfilment from the reward centres of the brain (in the hypothalamus), and someone else may feel only the uncomfortable chair and be worrying about the shopping list. While the spiritual side is important for some, others find great comfort in religious practice and attendance.

Studies show that for twins living at home, there is no clear genetic influence or difference from their parents in their practice. However, once they leave the nest, genes start to play a role.[10]

Elizabeth and Caroline were identical twins who came from an academic middle-class English family with an atheist father and agnostic mother. The sisters were very similar in appearance and character, both admitted to being stubborn, although Elizabeth was the naughtier of the two. At primary school they both became interested in Christianity and much to their father's surprise and displeasure they were baptised and prayed regularly. Their parents split up soon after and their father left home. They went through the normal teenage tantrums and slowly lost interest in organised religion and prayer.

After school they went to different universities. Caroline quickly rediscovered her faith; she became an even more committed Christian and joined student societies and church groups. Elizabeth began discussions with an Islamic group, initially arguing against religion, read the Quran to dismiss it and then found herself being drawn to and then converted to Islam. Both married and had two kids – Caroline with an English Anglican husband, and Elizabeth with a Pakistani Muslim (from then on she wore the veil – hijab – in public).

As she now says: 'I strongly believe that Islam is the one true faith and Christianity wrong. I endured many taunts and bigotry about my style of dress and beliefs and was often frightened to go outside. I once had to witness my three-year-old disabled son being spat at.' Caroline is similarly strongly opposed to her sister's Islamic views and 'her lack of belief in Jesus being the Messiah really upsets me.' She has had an easier time socially, but misses being close to her sister and having a drink with her. She says: 'I will never forget the fact that she very pointedly refused to sing hymns at my big day – a Christian wedding.' Both twins admit being saddened that neither could bear to act as guardian of the other's children because of their faith, although ironically they have much more in common genetically with each other's children than other aunts and share the same proportion of genes with them.

Sadly their mother, Annie, developed terminal metastatic lung cancer, which had the positive side-effect of briefly bringing the family back together. The closeness and bonding was short-lived. She admitted: 'I was initially bemused and then distressed by their fierce disagreements over faith, which being a self-confessed agnostic I just couldn't relate to. My main hope was to live long enough to see the birth of my two new grandchildren.' When, against the medical odds, she did, and was still alive nine months later, she had a revelation. 'I think I've found God,' she told her daughter Caroline as she recounted an epiphany moment she had while out walking. 'I felt his presence all around me – a spiritual presence. It's not just because I'm about to die – I'm not afraid of death. But I've changed my mind, there is more to life than just this current one.' She died shortly afterwards. Annie's genetic predisposition for faith, likely suppressed by her secular surroundings and her dominant atheist husband, may have been the crucial factor that influenced her daughters' uncompromising beliefs.

Where did this religious fervour come from? Neither had religious parents, and it is unlikely that the school alone could have had such an influence. Other twin studies have shown that after leaving home, children with the right predisposition can often switch religions, and that which form they then choose is not down to the genes but to life events or some mysterious unknown force.

Darwin and religion

Evolution, according to Darwin's theory, is the slow change over hundreds or thousands of generations of our gene pool through random gene variation and subsequent survival of the fittest. Traits and their gene variants that emerge by chance or selection, if useful, are kept, and disadvantageous ones are dropped. Religion or faith – like many other traits, particularly with a variable and genetic component – would have had to start somewhere. We tend to ignore religions that preceded monotheism 4,000 years ago. However, some eminent US geneticists like Francis Collins, who became a born-again Christian after reading the works of C. S. Lewis (and philosophers such as Thomas Aquinas), believe that humans have always possessed an innate moral law. Instinctively knowing right from wrong in every culture and tribe on the planet had to have come from somewhere, he argues. God and not evolution is the only logical source of this altruism and the universal search of humans for God.[11] Collins was the head of the US government's genome project (and now head of the NIH) and stood next to Bill Clinton in 2000 during the TV speeches. He refers to DNA, which is universal to all of earth's creatures, as 'The language of God'.

Others believe that this moral law is not innate but evolved, although there is considerable speculation about when and how this may have first occurred. Darwin (who originally tried his hand as a clergyman) himself speculated: 'As soon as the important faculties of the imagination, wonder and curiosity, together with some power of reasoning had become partially developed, Man would naturally have craved to understand what was passing around him and would have vaguely speculated on his own existence.'[12]

The fact that humans were receptive to religious thoughts

may be a by-product of our uniquely human trait of being able to predict the behaviour and minds of others. We will discuss this further later, but it was clearly a very useful trick, allowing early man (unlike other apes) to avoid conflicts and allow collaborations. Quite when in man's history this happened is unclear. Many link this development to the use of symbols. Figurines with half-human and half-lion bodies have been found dating back 30–35,000 years, which could indicate the existence of early shamans or witch-doctors.[13] Others suggest that religion might have evolved when societies got over-complex and collective decision making became difficult. To avoid repercussions to an aspiring leader if he got a decision wrong, he could avoid direct responsibility if he consulted the gods or planets or sacrificed some spare virgins.

David Dennett, a philosopher, suggests a theory for selection for religion.[14] Each small tribe in Palaeolithic times would have had some form of witch-doctor who for sick individuals was their only source of comfort or hope. If, as is likely, they underwent a form of hypnosis, on average this may have saved some lives – it still has a place in modern medicine. So any ancestors who survived will over generations have been selected to be more susceptible to forms of suggestion and persuasion. Anyone who has ever attended a public hypnosis show knows how a few people in the audience are instantly put under the hypnotist's spell. Thus a group of individuals susceptible to group beliefs that had modest survival traits may have arisen and continued to evolve, using religion as a protective emotional and collaborative survival trait over many generations. As groups grew in size, religion provided common goals and reduced selfish behaviour with the perception that 'a higher being' was always overseeing them.

E. O. Wilson, the father of socio-biology, proposed that many religions had clear health benefits – like the kosher laws of

Judaism that banned pork and shellfish and so reduced food poisoning. Although not all studies concur, there is overall evidence that religious people may even nowadays have subtle health benefits. One summary analysis of 40,000 people from 22 European countries showed that weekly religious attenders had half the rate of self-reported ill health of non-attenders.[15] Some of this benefit may have emanated from a sense of well-being, or optimism, or from the cultural benefits and social networks created. Others might say: 'It's God's hand.'

Another curious fact is that sufferers from a certain form of epilepsy, which affects the temporal lobes, often report that while having a seizure caused by tiny electrical impulses in this brain region they experience very firm religious beliefs and convictions. These include religious visions and voices, as well as having the urge to write down religious messages. This is understandable in medieval Europe and probably accounted for Joan of Arc's visions and fervour; what is remarkable is that it still occurs so frequently today. A recent survey showed about 10 per cent of sufferers in Brazil made the sign of the cross during their seizures.[16] As well as having religious DNA, could we also have parts of our brains formed for that purpose? Some studies have shown that enlargement of certain parts of the brain such as the middle temporal lobe are associated with greater religiosity[17] and others, like the precuneus in the centre of the brain, associated with scepticism.

Richard Dawkins, author of *The Selfish Gene*, coined the concept of 'memes' (based on the Greek word to imitate), which are the cultural equivalent of genes. The analogy is that genes are units of transmitted biological information and memes are units of transmitted cultural information. Thoughts, beliefs or ideas helpful to a 'meme's' survival can be passed on to future generations in a much faster way than painfully slow natural selection taking hundreds of generations. Examples include transmissions of melodies, fashion or ideas. Most successful religions also encourage the practice of procreation and large families, which jointly help both the religious genes and memes to spread rapidly – a process called cultural hitchhiking.[18]

Hollywood tales

There are plenty of examples of cultural trends that may be associated with certain receptive personality traits or genes. These include the relatively modern trend of women crossing their legs, or older recurring customs of piercing, tattoos or men's shaving customs. But an unrelated cultural trend that came to my notice was the recent fashion for shaving or waxing pubic hair. A medical friend mentioned that some patients were now apologising for not having waxed before a gynaecological exam, in the same way they might have done for having smelly socks. This habit is apparently mainly restricted to women aged under 40. Intrigued, I took a closer look and found surveys showing that the trend has hitchhiked and crossed oceans. This new 'religion' of the hairless pube has now taken firm root in the young.

A German study in 2009 of 18–25-year-olds found that 50 per cent of girls and 25 per cent of boys pruned to some extent. Recent large-sample studies in the US and Australia suggest that 20 per cent of young women are totally hairless.[19] Across Western countries, waxing rituals such as 'the Brazilian' (leaving a small remnant above the vulva (called 'le ticket metro' in France) and the 'Hollywood' (total) have become as common as having your nails done. How can a trend like this spread so quickly?

In the Netherlands, surveys have found a worrying subgroup of children who have developed 'pubo-phobia', with fears and obsessive behaviour about pubic cleanliness and hygiene, akin to some extreme religious intolerances. Some brave girls resisting the trends have been teased at school in the showers and ostracised. Some US psychology students agreed to do a trial of stopping body shaving for eight weeks and became traumatised by the social consequences. There is also one story of an American bachelor in his twenties who gallantly told the world that his older date, an aspiring female senator just under 40, was very sexy, until in bed he noticed that the waxing trend had passed her by. He promptly feigned tiredness and pretended to go to sleep. Clearly not conforming to a popular social trend can have downsides, and studies show the current cohort of depilators are more likely to have regular oral sex, be more adventurous and have fewer sexual problems.[20] How much of this is cause or effect is unclear, but it demonstrates how rapidly an idea can propagate when there is a perceived benefit, especially relating to sexual attraction.

So when did this waxing religion start? Were there any early prophets? Sources suggest it was unheard of in the West before the 1990s. Recent gruelling surveys by highly dedicated researchers who had to study 647 *Playboy* magazine centrefolds from 1953 onwards confirmed this trend toward the hairless 'Barbie look' that occurred from the year 2000 onwards.[21] The trend or meme likely arose about ten years before from an extremist sect in Hollywood, led by prophets like Larry Flint, called the porn industry. They continued it because it made stars look more 'impressive' and sold more films and so made money. Sociologists believe the combination of the recent reduction in female communal bathing and nudity and the easy access to the Internet in the 1990s allowed this meme to initially spread so rapidly by making women believe it was initially more common than it was in reality. Strangely, certain religious and geographic groups

– mainly in the Arab world and in Oriental Sephardi/Mizrachi Jewish female communities – have for thousands of years practised this. Some authorities believe the ancient Egyptian royal family, including Cleopatra, were proponents, but whether initially this was for social or hygiene reasons (pubic lice can be a nuisance in hot countries) was not recorded for posterity.

Jews and Muslims practise male circumcision and most religions prescribe a number of practices relating to beard and hair growing. These religious rules, although supposed to be God's commands, are a good way of identifying group members and stopping inter-faith marriages. The recent example of waxing our bodies shows how susceptible we are to following cultural practices relating to our hair, and how quickly trends can start and produce extreme proponents. Some of these same traits that make us conform so readily are clearly related to our susceptibility to religion.

Go forth and multiply – your genes

Most people when asked would guess that religious beliefs are universally declining, and the majority would say this is a negative thing. How is it that with these powerful evolutionary advantages, patterns are changing so rapidly over a few generations? A good example is Ireland, one of the most religious countries in the world. Surveys showed that 85 per cent of the population attended church at least monthly in 1988. By 2005 rates were down by a third, and still dropping due to recent sex-abuse scandals, despite Polish immigration. During this time, thanks to an 'economic boom' that has since gone sour, Ireland experienced an unprecedented doubling of per capita gross domestic product. As consumerism increased (spending on holidays, leisure and alcohol in particular), so did secularism – switching from worshipping God to goods.[22] While some of these lapsed church non-attenders might still have had faith, beliefs also fell over the same time period. In a land where not long ago it was the most popular and prestigious profession for many, priests are now risking extinction. In Dublin diocese there are 45 priests over 80 but only two under 45, and the whole country is predicted to lose two-thirds of its priests within 20 years.[23]

In most of Western Europe, where base rates were already much lower than Ireland, churches are being converted into apartments, priests and nuns can't be recruited, and believers are now in the minority. In the UK the numbers of believers have reduced from 64 per cent in 1991 to 48 per cent in 2008 – a drop of around 1 per cent per year.[24] European cultures are becoming increasingly secular, with organised religions slowly losing influence and power. So in the modern world, is the decline of religion now inevitable?

In some countries a natural experiment occurred where the effects on religion of sudden culture changes other than economic booms could be measured. In Russia, where nearly all the population believed in God at the start of the 1917 revolution, religion was heavily suppressed for 70 years. Just after the end of communism in 1991, the number of believers was 22 per cent of the population, with a third of previous non-believers under the old communist regime recently converting.[25] Most were young Russian males. These 'revivalists' reported this to be a miraculous turning point, one that suddenly gave a purpose to life.

Compared with East Germans in a similar situation, Russians, despite similar baseline rates of belief, had four times the recent religious conversion rates. They are also more enthusiastic church attendees, with more extreme socio-religious views – the majority believe that all anti-religious books should be banned and only religious figures should hold public office. Socio-economic factors may have mellowed East Germans. Elsewhere in Central and Eastern Europe post-socialist increases were seen in places like Hungary, where regular church attendance has tripled in ten years. Unlike the changes in Europe and many other countries in the last few decades, religious beliefs and practices in the US have remained firm and stable, despite its affluence and materialism.

These national figures are averages and disguise the fact that within every country some people of religious backgrounds are losing faith and a proportion with no previous religious background are taking it up. So even if not always expressed outwardly as 'religion' or belief in God, these faith-susceptibility genes of our ancestors can't disappear from our genetic make-up so easily and must still be an influence in many of us.

We know that religious people are encouraged to have more children. Recent surveys of 82 countries in the World Values Survey 1984 to 2004 found that women attending some kind of weekly religious ceremony had 2.5 kids on average and those not attending had only 1.67. With stricter orthodox religions the results are even more striking – old order Amish couples have an average of 6.2 children, and most other strict religious groups have a fertility at least three times higher than the average woman.[26] However, some raised within strictly religious communities are tempted to leave and join the secular world. Recent predictions of future religious gene rates in the US population, balancing fertility versus desertion in each generation (through cultural hitchhiking), have given surprising results.[27] The genes of the smaller religious groups would not only continue to survive, they would actually grow and make up an increasing percentage of the population. If a group like the Amish or Orthodox Jews with three times the fertility rate made up 0.5 per cent of the population and lost 5 per cent of its members every generation, its genes would account for 20 per cent of the population of the US after only ten generations. Even if 50 per cent of the group defected and became secular, it would still dominate the US after 20 generations.

So even in secular societies like those of Western Europe, while overall fertility rates are dropping, the genes would infiltrate into the main population due to slow dilution from the high-fertility religious groups. In the UK the Muslim population has tripled in 25 years and has recently reached 2.5 million, due only in part to immigration.[28] Globally, the Muslim population is forecast to grow at about twice the rate of non-Muslims over the next two decades – an average annual growth rate of 1.5 per cent for Muslims. However, it is expected to then slow down due to changes in female education and prosperity. The Amish population in the USA has already doubled in twenty years to 250,000 in 2010 and is predicted to increase exponentially to reach 40 million by 2050. Changes in other orthodox or extreme religious groups (Hutterites, Othodox Jews, fundamentalist Muslims, Mormons etc.) are likely to be similar, stimulating a possible economic revival in traditional black cloth and hats, and beard-grooming equipment, as well as more eco-friendly horse-drawn carts and facilities for manure clearance.

Gene deserts

Scientists have claimed to have found the God gene in the past. This was based on the old methods of linkage and candidate genes which implicated the VMAT2 gene – a brain neurotransporter. Sadly like many 'discoveries' of the early genetic era before 2007, this turned out to be a false dawn.[29]

My group, too, have been trying to track down the genes

responsible for religious belief (or disbelief), using the modern analytic methods of half a million DNA markers (genome-wide scans), and are getting really close. Using 4,000 of our UK twins we narrowed the search to a clear signal from a small stretch of DNA on chromosome 15. This could only occur by chance 1 in a million times. When we looked to see what gene this signal was close to on the chromosome, we had a surprise: there was no gene anywhere near this marker. This area is known by geneticists as a gene desert, because it is huge and was believed to contain nothing useful, being full of what was known as junk. Mysteriously, quite a few real associations with diseases (about 1 in 4) end up in these desert areas – something that we can't yet properly explain.

While it is just a matter of time before gene variants like these, which alter your chances of divine belief, are convincingly confirmed in humans, their origin will cause major disagreement. Atheists will probably argue that they have been selected for by the randomness of trial and error – i.e. natural selection over many generations – and are evidence for the non-existence of God. Believers in a greater power will no doubt counter that the presence or absence of genes like these is part of a creator's perfect master plan via God's language, DNA, allowing faith and inherent morality to occur. The scientist mainly responsible for the split between science and the Christian Church doctrine – Darwin – himself was clearly torn in his beliefs. He expressed 'the extreme difficulty in or rather impossibility of conceiving this immense and wonderful universe, including man with his capacity of looking far backwards and far into futurity, as the result of blind chance or necessity. When thus reflecting I feel compelled to look to a First Cause having an intelligent mind in some degree analogous to that of man; and I deserve to be called a Theist.'

Losing faith

Alice and Sally were identical twins born in 1948, brought up together in Lambeth, London, close to my hospital, in a working-class Christian home. While both of their parents believed in God, they didn't go regularly to church, but their two shy and pretty daughters attended Sunday school, which they enjoyed and which played a key part in their social lives. They were baptised and confirmed and both twins prayed regularly. Although in different classes at school they shared a room and were otherwise always together. They sometimes had violent fights, and were always called 'the twins'. After they gained good final 'A' levels at school, their working-class parents didn't allow them to go to university, thinking it 'a waste of time'. They both got comfortable jobs in different branches of the Civil Service. Then their paths began to diverge.

When she was 20, Sally married a self-employed builder with family money. She continued to work for six years, but despite the financial security was not happy. After four children and several affairs she decided to abandon the marriage and divorced. She remarried soon after and is now happy with an older man who resembles her father. Alice, in contrast, had an early broken engagement and decided to wait for Mr Right. Finally at 35 she married a much older man, and stayed faithful. When their father, whom they both adored, died of lung disease five years later they were both devastated, but their religious paths changed dramatically. Sally lost her previously firm belief: 'If God existed – he wouldn't have let my father die. I no longer believed in God, an afterlife or the point of church.'

Alice continued to believe and said that she was shocked by her sister's reaction: 'God doesn't work in that way. He helped me come to terms with my father's death and I know I will meet him again in Heaven.' Her father's death actually strengthened her belief: she regularly attended Catholic mass and prayed daily. Alice doesn't think she is a very spiritual person, she just believes firmly that God is there and loves her. Sally also doesn't believe she is very spiritual, but she still remembers the comfort she used to get from praying. Sally is generous, gives to many charities and has very strong views on the planet and global warming, being a member of Greenpeace and the Campaign for Nuclear Disarmament. Her views on helping the most needy and strong socialist principles are in contrast to her religious sister Alice, who firmly believes that most people should learn to stand on their own feet and compassion and charity should be reserved for those that try.

The contrasting religious twin stories provide an insight into how our beliefs are shaped. While all three sets of identical twins probably carry the religious susceptibility genes, they behaved and expressed them very differently. Are beliefs simply a mix of upbringing and genes, as the first pair of twins, Debbie and Sharon, suggest, where genes and culture work in tandem? Or can small differences in life events change a Christian to a Muslim, as in the case of Elizabeth and Caroline, or – as with Alice and Sally – a bereavement change something so fundamentally human as belief or disbelief in a universal creator or afterlife? These twin stories and the changes in populations over time suggest that our thoughts, beliefs and practices are less fixed and much more malleable than superficially they appear. It suggests that even our patterns of beliefs could be altered epigenetically – with faith genes switched on or off.

If so, could we modify the behaviour of, for example, religious fundamentalists as easily as our missionary ancestors converted non-believers? Can materialism and prosperity mixed with a sense of security reduce faith? Or could we alternatively have another resurgence of religion in the West, as suggested by the data from Russia, where dormant genes may be switched on? These faith genes are still very much present somewhere in our genomes and may be actually growing in the population, so in a secular world they could be influencing our behaviour in new ways that we are unaware of. This could take any number of forms.

People who are susceptible to faith are more likely to also have conservative views, with greater respect for tradition and authority.[30] In the UK surveys show that religion is not associated consistently with any particular politics, but religious people are more likely to vote than non-believers, suggesting that subconsciously they wish more strongly to identify with other groups. In the absence of religious culture (as in the old Soviet Union or secular Europe), others might seek or find comfort in other authoritarian ways. Could the growing number of small groups of individuals with extreme inflexible views on the world but who can't express themselves through religion come into this domain? Surveys in the US suggest that beliefs in the paranormal are currently on the rise, with 54 per cent of Americans believing in psychic healing and 50 per cent in ESP.[31] The findings are also true of US students.[32] Those with the strong beliefs may end up identifying with other like-minded people. This could include extreme vegans and caloric restrictors, obsessive gym worshippers, extreme political groups, believers in alternative medicine, Apple techno addicts, AIDs and vaccination deniers, conspiracy theorists, or believers in alien abductions.

It is also possible that extreme atheists – those who are trying to convert believers and are less than tolerant of religions – could also be carrying these faith genes, though highly unlikely to admit it. Ancient philosophers and more recently psychologists suggest that it is natural for humans to question their own existence and seek higher powers. Yet we are still surprised when children like Helen Keller – totally deaf and blind from 19 months of age and isolated – can still ask unprompted: 'Who made the sky, the sea, everything?' Richard Dawkins, when asked in an interview what he thought our purpose in life was, replied: 'It's not a proper question to put, it doesn't deserve an answer.'[33]

Most of us, whatever our genes, would probably disagree.

5

'THE PARENTING GENE'

Nature, nurture and naughtiness

The baby was delivered by a self-taught midwife who immediately covered and rubbed her in salt to get rid of the sticky birthing liquid. She was wrapped in bandages five layers deep round her body and her head was fixed onto a stiff board. The baby was first purged of its 'long-hoarded' excrement by castor oil or enemas and then given a strengthening glass of wine or whisky. The baby would be put to the breast only on the second or third day, when the mother's milk was ready, or if the family was rich a replacement mother (a wet nurse) would be paid. If none could be found or there were some health concerns (such as, commonly, syphilis in wet nurses), the babies would suckle direct from goats' or asses' teats. The content baby was swaddled up all day so she couldn't move. She was hung on a hook, usually in the kitchen, and brought down to feed.

This is not a story of a modern organic New Age birth in north London, but how most infants were treated in England and France in the first few weeks of life in the seventeenth or eighteenth century.[1] We have less detail about what happened before then, but what we do know is that childhood was not a protected or isolated event. Children were just treated as small adults.[2] With no real concept of education for most families, children were exchanged after the age of six into apprenticeships to learn a trade or serve a noble. As for the pre-school age, the child was usually a nameless piece of livestock and roamed the townships in street gangs, wore rough clothes, rummaged tips for food and contributed to the family's income with petty theft and begging. Parents couldn't afford to involve themselves too emotionally. These rough childhoods did not necessarily end in disaster though. The future Emperor Frederick II (1194–1250), heir to the most powerful dynasty of his time, the Holy Roman Empire, and one of the best-educated and most enlightened rulers in history, who was fluent in six languages, had passed his early childhood and adolescence as a thieving thug in a Sicilian street gang.

History tells us that life for an infant or adolescent 300 years ago was certainly very different from today. There wouldn't have been much talk of early bonding or the crucial importance of listening to Mozart. Perhaps because – even up to the nineteenth century – only 50 per cent of children survived past the age of six, and quite a few mothers died in childbirth, emotional bonding with children was seen as a lesser priority. So is parental concern for children's upbringing a new phenomenon invented by us and psychologists in the 1960s and 1970s?

Indeed the word 'parenting' itself is a neologism that only appeared in the Oxford English Dictionary in the 1970s. It may be new but it has been marketed extremely well and is today a global industry worth billions. The best-selling parenting book of all time, by Dr Benjamin Spock, has sold 50 million copies across four decades and is outsold only by the Bible.

Yet, if it is only in the last three generations that we have been taking parenting from the child's perspective seriously, why have rates for nearly all behavioural and psychiatric disorders in children steadily risen since the 1960s? Why are major expensive programmes in the US and UK trying to redress this and teach good parenting techniques? What is going wrong, and what techniques could put it right?

Parents and guilt

I was asked recently by the BBC to discuss a new book by the American economist Bryan Caplan called *Selfish Reasons to Have More Kids*.[3] In it he argues that the scientific evidence suggests that parents shouldn't feel guilty. His message is that you can't change your kids by trying too hard. If they don't want to go to piano or sports practice, relax. If they don't want to read books now, they will later. So take it easy and enjoy them without stressing. This is completely the opposite advice to that given by Amy Chua, the author of the best-selling book *Battle Hymn of the Tiger Mother*.[4] As well as being a Harvard law professor she dedicated her life to her kids to make them perfect Grade A students and musical prodigies. She wouldn't allow them to get B grades, go to sleepovers, waste time in school plays or to not practise for up to six hours a day – even on birthdays, or holidays. She believed that Western attitudes are too soft and the global success of Asian kids is due to the firm maternal approach.

She did succeed in her goals. Both her daughters were straight-A students, and musical prodigies, and both played at Carnegie Hall. However, the younger feisty one, after many years of fighting and punishments, at age 13 eventually rebelled, gave up the violin and played mediocre tennis instead. The other daughter was quieter and meeker and continued piano practice, although her mum later found secret bite marks on the piano that had been there for years. Apart from the occasional moment at the end of a prestigious public concert, joy and happiness in mother or daughters don't feature highly in the book or in their lives.

Time will tell whether these kids are grateful to their mother and raise their kids the same way. Yet you might wonder whether the attractive healthy daughters of two high-achieving Harvard law professors would not have pretty good genes and a reasonable start in life without their mother's excessive and obsessional devotion. Nonetheless, books on parenting like these by a variety of psychologists, journalists, economists, mothers, midwives, doctors and nurses sell every year to an eager market of insecure parents trying desperately to do the right thing.

The parenting debate is far from new. Books started in the eighteenth century when the philosopher John Locke (who like quite a few experts when they wrote their books had no children) argued for firm regimens to harden minds and bodies for the future for the good of society. These would include irregular mealtimes, putting the children to sleep anywhere, often wearing cold or damp rough clothes in poor weather, and educating primarily by fear and awe. This boot-camp style of parenting was countered by Jean-Jacques Rousseau in France, who argued the complete opposite – that nature should be allowed to take its course, children should be allowed to have as much attention and breast as they wanted, roam free and feed and sleep at will, and have no schooling, discipline or training until the age of 12. Interestingly, Rousseau had 5 children and was not the ideal Dad, raising them 'naturally' in an orphanage. 'Rousseau fever' took off in a big way; the English politician Charles James Fox was famous for being raised 'naturally' and as an 18-year-old used to climb on the table at dinner parties. Jacob Abbot famously remarked that 'had Byron and Washington swapped cradles, the world would have been very different'.

Although the concept of the importance of a firm hand, fresh air and regular cold baths with no cuddles slowly prevailed, there was a continual swing of opinion every few decades and wide variations in practice. These included heated debates about breastfeeding, wet nurses, stimulation, discipline, potty training, the role of the father, and use of nannies and grandparents up to 1945.

After the Second World War, in 1946, Benjamin Spock's book came out and opened the floodgates of international baby-rearing best-sellers. He was a New York paediatrician born in 1903, influenced by Freud and the attachment work of Bowlby and the major social changes. He reflected a popular view as a reaction away from the rigid authoritarian parenting styles of Nazi and communist states. His general approach was: listen carefully to the needs of your child, let them sleep when they want, don't be rigid about mealtimes, and think why they may be crying.

Spock's books were full of common sense and confidence, but the message was clear: fail to respond to your baby correctly and it has consequences. That message, coinciding with an increase in our leisure time, was when our current era of parental guilt probably began. Spock was on most families' bookshelves in the US and UK for 40 years, with new editions every ten years reflecting considerable changes each time, such as views on breastfeeding and working mums, until the last edition just before he died aged 94 in 1998.[5]

For the last 30 years Gina Ford has been one of the most widely read parenting authors, selling nearly 2 million books. She is a former midwife who favours a strict routine-based system about meals, feeding and education, and discipline more akin to the hardy pre-war era, without mentioning bonding.[6] Ford has been called the Delia Smith of parenting, with a recipe for all situations. Her nearest publishing rivals are Dr Bill Sears and his wife Martha, who believe in a different approach, namely the crucial importance of attachment and one-to-one maternal bonding above all else.[7] Today some of the recent debates are on the role and responsibility of the mother, with many – such as the Searses in the USA and in the UK psychologists like Oliver James or Penelope Leach – increasing the guilt by telling working mothers that nurseries will damage their children, crying causes brain damage, the Gina Ford routines are bad, using the naughty step time-out disciplinary method is counterproductive and that toddler misbehaviour is all their fault.[8]

As historians of childhood reveal, none of these debates or conflicting theories are really new, and although the self-proclaimed experts can't agree among themselves on which parental advice is best, they all have one thing in common: they are based on hypotheses or observations that fail to control for genetics and individuality. They also want to sell books, and guilt and controversy work best for this. In his book *A Good Enough Parent*, Bruno Bettelheim says that 'parenthood is an art accessible to any human being, not a skill learnt by listening to child-rearing experts'. Many of these experts, he argues, come from broken homes and have their own issues.[9] 'Scientific observations about babies are more like mirrors that reflect back the preoccupations and visions of those that study them.'

Amid the lucrative bickering over the different styles of parenting since the war, a few lonely dissenting voices suggested that both groups had got it wrong: parental styles just didn't matter. Bryan Caplan was echoing the work of the self-taught psychologist Judith Rich Harris, who had earlier written the first major book (*The Nurture Assumption*) debunking the behaviourist claim that family environment and parental style were crucial.[10] Her ideas came from the failure of multiple twin and adoption studies to find any significant effect of the family environment on personality.

To further explore the real implications of the family environment, and how all this relates to genes, let's look at a story of twins brought up in very different surroundings.

Nina was born just after the war and was eight when she was first told she had a twin sister who lived a few miles away. Before then she had been brought up as an only child by a loving couple who were in their late fifties and had no kids of their own. They treated her well and spoilt her with presents and holidays. Her dad used to read books to her on his lap every night and always had plenty of time for her, as he had retired early. She was often naughty, but was never smacked or badly scolded and ended up bossing her poor mum around. She had the run of a large house and garden.

Despite being quite lazy, Nina went to a good local school and passed the 11-plus entrance exams at the second sitting to go to a grammar school, and then university, where she studied to be a teacher. 'I always had everything I needed, although sometimes wondered what it would have been like if I'd been born into a large family with younger parents or kids to play with.' She ended up marrying another teacher and having three kids, all of whom also went to university. While Nina was young she was discouraged from seeing her sister Gill by her adopted parents, who feared she would return to her biological family. So apart from a few fleeting visits they never really got to know each other until they were in their forties, when they were flown to Minnesota for batteries of tests as part of the adopted-twin study.

Gill had a rather different life. She was never alone. She grew up as the youngest of six kids and the family of eight were squashed into a tiny terraced house with a toilet that was across the road. 'I can never remember having a meal sitting at the table, as there were not enough chairs, although I don't think we ever went hungry. I don't have any fond memories of childhood apart from the highlight of once being washed on the kitchen table. When I was naughty – which was often – I was smacked and told off and sent to play in the street, where I remember spending much of my time'.

Her father was pleasant enough, but invisible, as he worked 16 hours a day as a train driver to pay off his gambling debts. Her mother was 'a bitter woman, totally indifferent to us, other than providing us with food'. She had tried to give both twins away at birth, but one unofficial adoption arrangement (Gill) failed. When Gill came first in her class at school, nobody cared. 'My report was left unopened on the mantelpiece.' One of her older brothers was actually illiterate. She didn't receive much attention at the local school either, which had a poor reputation, and she failed her 11-plus exam. She left school at 14 with her friends to work in a shop, but got pregnant soon after and married a trainee electrician. Life didn't get any easier then, as he was a regular wife-beater, often drunk, and on one occasion he cut her with a knife. They stayed together on and off with further violence and poverty for 14 years and two more children. Gill did eventually return to school, aged 29, to get some O-level qualifications, and then got a good job in the local Civil Service, divorced her husband and remarried a gentle kind man. They are now retired and spend half the year sunning themselves in the south of Spain.

Talking to Nina and Gill now, they seem much the same, although Gill, who smoked for most of her life, admits that she had a tougher life and, unlike Nina, has lost her teeth, and has poorer health. I asked them both if they would swap their lives. Nina said that although she would have liked to have known a family, she wouldn't have changed anything. Gill, although now happy, said after some reflection: 'Life would have been a lot easier for me if I would have swapped.' One surprising detail that Gill told me at the end of our conversation was that at age 16, just like her sister (but unknown to her), she also fell in love with a trainee carpenter, got pregnant, and married at 17. Like her sister she had fallen for another aggressive abusive male, but unlike Gill she left him after only a few weeks, with the support of her family. Gill, without support from her family and no independent income or home, had had to put up with her abusive husband for years. This clearly changed her life. Both twins believe they are very similar, with the usual identical mannerisms and similar personality. They agree that Nina is the more diplomatic, and Gill speaks her mind, but both try to be assertive in different ways. They believe they raised their kids pretty much the same way, and the next generation now see each other regularly.

So where does this example help us with the parental debate? Would social workers nowadays (backed up by Article 19 of the UN Convention on the Rights of the Child) have moved in to relocate Gill from her overcrowded poverty-stricken home and cold-hearted mother who had wanted to give her away? Figures show that when a child develops antisocial behaviour (conduct disorders) the state and taxpayers pay a heavy price. In the UK alone each child with problems costs around £70,000 directly and up to £500,000 indirectly, with a total cost of children in care of around £2 billion. Advocates of parental training estimate that it costs only around £600 per family. Surely this is good value if it works?

Clearly the two girls had very different family environments and parenting styles. Yet the differences – absent father, cold tough mother, etc. – didn't stop their naughty tendencies or both falling to the temptations of lust and getting pregnant at 16. It did not seem to change their personalities markedly in the long term, although it clearly affected their education, confidence and finances, and ultimately their route to happiness – which for Gill took much longer to reach.

Why are children so different?

My colleague Robert Plomin, who works on young twins, asked way back in 1987 the key question: 'Why are children in the same family so different from one another?'[11] He was referring to the fact that normal siblings should, judging by their similar genes, in theory be more similar than they are, and adopted siblings never seem to grow more alike, even when raised in the same family all their life. This raised the question of the real importance of the environment. Now environment is traditionally divided into what is unique to an individual or random (such as being mugged, or having an accident) and what is common to the family or shared by both siblings (such as housing conditions). Behavioural geneticists were keen to work out how much the shared family environment (which includes parenting) really influences future personality. To do this they performed over 43 twin and adoption studies before 2000, just on this question alone. Many of these studies were small and poor-quality, but pool them together in a so-called meta-analysis and clear patterns emerge.

The researcher Eric Turkheimer[12] found in summing all the data from the 43 studies that a meagre 2 per cent of differences in behaviour were due to parental influence, the same as estimated for sibling interactions. Even less important was the 1 per cent effect estimated for birth order and age effects. This is contrary to many psychologists' firm belief in the power of birth order on personality. As usual genetic factors explained about 50 per cent and random and specific environmental effects the rest.

Because of the controversy and strong views on both sides of the debate, a large study (called NEAD – non-shared environment in adolescent development) was funded by the

US government to answer the question decisively. It brought together sceptics from all camps to design and run it and didn't exclusively use twins. Starting in 1988, they observed 720 pairs of different kinds of early adolescent siblings in stable two-parent families, which were not easy to find. The pairs included identical twins, fraternal twins, ordinary siblings, half-brothers and -sisters and step-siblings. They followed them intensively over three years and again after 11 years, by interviewing them, their parents and their friends and peers. It was a huge task – the most thorough examination of the family's influence yet performed – and most expected it to confirm the importance of parenting.[13]

One of the researchers, David Reiss, was a psychiatrist and family therapist, and before the study a keen advocate of the parental influence. He was 'shocked by the results'. It showed as expected a large effect of genes on outcomes and behaviour. But crucially, after accounting for genes no effect was seen of distinct child-specific environmental factors in parenting, or parental biases in the way the kids were treated. Nor did the study show evidence of identical twins being treated more similarly. In fact the researchers were unable to explain any of the presumably huge influence of unexplained non-genetic factors on behaviour at all. This was a clear blow for the nurture camp and led to the first public questioning of the real role and extent of nurture in parenting, which had hitherto been untouchable.[14]

The NEAD study confirmed that there was a correlation between parental handling of adolescents – i.e. harsh or lax, loving or indifferent – and later adolescent behaviour. However the underlying reason for this was genetic, both in the mother's genes and the child's. Furthermore the same genetic factors that influenced maternal harshness also influenced adolescent antisocial behaviour. Just over 70 per cent of the agreement between maternal treatment of children and their antisocial behaviour was accounted for by these common, but still unidentified, genes.[15] In younger pairs of US twins and siblings followed through to adolescence, only a weak correlation was found between early maltreatment and subsequent behaviours. Moreover, the greatest effect seen was not the reaction of the kids to their parents, but the parents' negative response to the genes of their poorly behaved kids.[16]

This idea that there are important individual responses of the parent to each child, reacting to their genes, is a crucial one. It is not so much the parent who influences the child, but more the child's genes that influence the parental response.

But environment still has a small but important role to play. Re-analysis of the NEAD study showed for example that up to 5 per cent of the environmental effect on adolescent depression was due to maternal negativity.[17] But in general, environmental effects were very specific to the child and not usually shared by the same siblings or twins in the family.[18] This is odd, and suggests that even identical twins can react very differently to the same environment.

To support this idea, a study of identical twins in the same classrooms found they did just that. They reported quite different responses (both positive and negative) to exactly the same teachers and same lessons given at the same time. In this study Robert Plomin's team followed 122 ten-year-old identical twins with daily diaries at school for two weeks.[19] They found overall only a modest 65 per cent agreement on perception of classroom experience. The twin receiving the more positive vibes about school from teachers and peers had increased success, particularly in maths and science, by 8–15 per cent. Clearly small differences in perceived 'positivity' can make quite a large difference to learning.

The fact that even identical twins can experience the same environments differently reinforces the observation that generally siblings from the same family and upbringing turn out more different than we would expect, given that they share half their genes and most of their environment for the first 18 years. It is almost as if we all have a built-in individuality processor working all our lives to make us more different to our siblings than nature programmed us.

Neglect and resilience

For the first 30 months of her life, Simona languished in a Romanian orphanage. She lay in a cot alone for up to 20 hours a day, sucking nourishment from cold bottles propped over her tiny body. Unable to sit up by herself, she would push her torso up on stick-like arms and rock back and forth for hours, trying to soothe the aching void that had replaced her mother. Like the other babies, her head was shaved to reduce infections. She shared a large room with 24 other cots, mostly in the dark. The difficult older kids were tied to the side of the cot. On the other side of the room was the oldest child, six-year-old Stefan, who looked half his age, his eyes wide and brown. He lay mostly on his back, occasionally managing to pull himself up to peer out of his little prison. Until a few months ago, when an American volunteer visited, no one had bothered during his six years of institutional life to encourage him to sit or stand or speak.

As if that wasn't bad enough, 10 per cent of the kids had HIV, due to experiments by the incompetent health minister Dr Iulian Mincu, who authorised micro-blood infusions in a hopeless bid to boost the undernourished kids. Simona and Stefan were in just one of 35 state-run orphanages in Bucharest. After the fall of the Ceausescu communist regime in 1989, the

West was able to see the wider legacy of his policy. In 1966, a year after Nicolae Ceausescu came to power, Romanian State Decree No. 770 declared abortion illegal for any woman under 45 who had not yet produced four children. In 1989, in pursuit of a larger workforce, this was increased to five. Birth control was virtually unavailable, except for those with access to the black market. The result, because of the poverty and lack of space: an orphan population of over 120,000.

When Western doctors entered the orphanages a few years later they were shocked not only at the filth and health of the kids but at the standards of care and apathy of the staff. They were casually chatting, laughing and smoking in the corridors while the babies literally rotted in their rows of cots behind closed doors. They estimated that each baby got only about six minutes of stimulation per day. Toys were not allowed, as the few they had caused fights between the older children and so were only brought out for visitors. Sadly the situation didn't improve much under Ceausescu's successor Iliescu, who unbelievably kept the same incompetent health minister and the same ideas about the strong role of the state in childcare. Some of this apathy may have been due to the majority of the children being Gypsies who in half of cases were retrieved by their families after the age of three, when they could start to be useful. So can we predict what would happen to these three-year-olds after this experience of extreme maternal deprivation?

The children still in Romanian orphanages 'look frighteningly like Harlow's monkeys', said Dr Mary Carlson, at the time a neuroscientist at Harvard University, referring to a well-known experiment of the late 1950s in which baby monkeys were removed from their mothers a few hours after birth and reared without parental care.[20] The infants developed abnormal habits and social behaviours. They were tiny for their ages and often sat and stared for long periods, or would rock back and forth for hours.

The ex-street urchin Frederick II, Holy Roman Emperor, performed an earlier environmental experiment in the thirteenth century. He raised dozens of children in silence in an attempt to discover the natural 'language of God'. It was not a great success, unless God's preferred language was silence. The children never spoke any language and apparently all died in childhood.[21] Other anecdotes exist of nature showing that stimulation is necessary for normal function. If a young chick has its eye covered from birth and the bandage taken off after six months, the brain connections are never made and the chicken remains blind.

In a few short years, one single cell, the fertilised egg, becomes a walking, talking, learning, loving and thinking being. In each of the billions of cells in the body, a single set of genes has been expressed in millions of different combinations with precise timing. Development is a breathtaking orchestration of precision micro-engineering that results in a human. In order to create the brain, a small set of precursor cells must divide, specialise, connect and create specialised neural networks involving trillions of connections. It's not surprising that this process requires nourishment. The first three years are now known to be the most crucial for normal development. All this complexity and Emperor Frederick's experiments suggested that the orphans deprived of attachment, emotion and sensory input would never recover.

One of the largest studies following up Romanian orphans is the English and Romanian Adoption Study (ERA) by Michael Rutter and his team, down the road from me at the Institute of Psychiatry.[22] It has been known for a long time that profound deprivation in early childhood increases the risk of subsequent psychopathology. However, the different types and range of problems were unclear. They followed up 165 adoptees who had spent more than two years in these terrible institutions, usually from birth to three years, and compared them with UK adopted infants.

When they assessed behaviours at follow-up for up to 15 years, although there were a few severely disturbed children, most had improved. However, on average they still showed persistent problems with attachment, inattention, overactivity and autistic-like behaviour. The study suggested that the first six months of deprivation were the most crucial, with longer durations of misery strangely not making them any worse. Most of the major physical problems and most of the cognitive deficits had actually resolved fully with good nutrition and stimulation. Oddly they found no significant increases in the risk of other problems such as emotional difficulties, peer relationships or conduct problems. Also 20 per cent were found at age four to 15 to be completely unaffected in any domain whatsoever – suggesting that some children were incredibly resistant to the most appalling of conditions,[23] and that environmental determinism is no more absolute than genetic. The data pose many questions. Why should some children be affected so much more than others? Why could some stay normal? Other studies found that the brains of these deprived children showed abnormal metabolism based on PET (positron emission tomography) scans in many parts of the brain responsible for emotions and empathy. But why would only some parts of the brain and some emotions be damaged?[24]

Lickers, groomers and bullies

Clearly one simple way of explaining differences in response between kids is differences in their underlying genetic backgrounds. The group found some evidence that dopamine genes were also involved in explaining who was worst affected by inattention disorders.[25] But while this may explain some variation, it isn't the whole story. In the longterm study of normal kids in the Dunedin Study in New Zealand that we discuss more later, those kids with different versions of the genes controlling the neurotransmitters (particularly dopamine and serotonin receptors) react differently to the same bad stimulus.

So clearly some level of interaction between our genes and environment is happening, but how? Could animal studies help us understand the mechanisms by which genes could be influenced or modified? The early classic Harlow monkey studies, taught to every psychology student, that I mentioned earlier showed that infant monkeys separated early from their mothers do badly. But they couldn't properly separate the effects of genes from environment, as monkey mums are not always very keen on adopting other mums' difficult offspring.[26]

However, clever rat experiments by Mike Meaney's lab in Montreal have managed to partially and more realistically reproduce the human experience in a way that crucially can account for genetic differences. When identical strains of rats are separated into two adopted groups, one of them raised by adopted mothers who are naturally attentive 'good mums' and are called high lickers and groomers, the other by 'cold mums' who are naturally low lickers or groomers, major changes in their behaviour occur. The pups who received less licking and grooming in the first week develop significantly greater levels of stress and anxiety.[27] The team found that a key brain hormone was involved, called the glucocorticoid receptor, a steroid hormone regulator that when blocked causes levels of the stress hormone cortisol to rise in the hippocampus and other brain areas of the pups. This meant that these underlicked pups would react more to some perceived threat and be in a state of greater alertness.

The same group has shown that modification of the key regions of genes like the glucocorticoid receptor can affect the function of literally hundreds of other genes downstream, causing a cascade of effects.[28] This means that if the key brain genes are switched on or off epigenetically, it will have widespread effects. Some of the biggest gene effects are seen in gene families (called protocadherins) which are involved in making the trillions of brain synapses (connections in the brain). These changes could be reversed by cross-fostering with an attentive licking mother.

Other key hormones being influenced by licking and grooming were oxytocin and dopamine. Oxytocin, known as the cuddle hormone, is released during suckling and is key to how animals and probably humans form bonds with kin and lovers.[29] It activates certain brain areas.[30] Importantly, the changes in the oxytocin gene were also transferred to the daughters, although they were raised normally, and influenced their behaviour with their own pups.[31] These changes could have some evolutionary benefit in other animals as well as rats. Making animals alert and anxious can actually be helpful in some situations and may increase the survival or mating success of the pup in stressful or dangerous conditions.[32]

So the combination of history lessons about parenting and the huge variations of styles over the planet and in time, plus twin and adoption studies, seems to be clearly telling us that for 95 per cent of families parental style and shared home environment has little to do with how your kids turn out personality-wise. But this leaves several questions unanswered. If genetics explains around half of the differences between kids and parental styles don't explain much else, then what is explaining the rest? Why do siblings in particular end up being so different and respond differently?

Bullying in schools is a very common trauma, with around 1 in 6 kids admitting to being a bully and the same number to being a victim. Although the traits of both bully and victim are highly heritable, context and family support are also important. When both identical twins are bullied at school, the one that perceives or actually receives greater warmth from the mother will experience less trauma and may become more resilient.[33] When for example only one identical twin experiences bullying, studies have shown that their response to future stress is permanently reset as compared with the un-bullied twin.[34] So it is clear that being a sympathetic parent that the child can come to for support is still an important feature, even if its effects may be hard to quantify.

To go back to my original question – do parents matter? – clearly the answer is still yes – but the extent is unresolved. Kids without a source of emotional attachment on average do very badly in life. Those who are neglected by their parents and with no sense of security have a high chance of being disturbed or suffering social problems. But the level to which normal ranges of parenting in most families affects children has undoubtedly been exaggerated. Twin studies have probably gone to the other extreme and underestimated environmental effects on the individual.[35] This is partly because these effects are not as predictable as believed and there are few shared family effects.[36] There is however no good scientific evidence that for example being a part-time or full-time working mother, duration of breastfeeding, being an active father, levels of strictness, number of cuddles, bedtimes, mealtime routines, TV watching, supervised home reading or homework supervision make any long-term differences in your child's development or ultimate personality or behaviour.

All parents like to think they have had a role in shaping their children – it's part of being a parent. So by all means let us take the credit when it goes well. But if things don't go quite as planned and your kids aren't as perfect as your irritating friends with perfect A grades, impeccable manners, piano prodigies and tidy rooms, don't blame yourself or your genes – blame your kids' genes. We have seen how children's genes will drive them to seek more or less of the environment that surrounds them, and they will be drawn to what they like. Their genes can also drive you crazy. Their genes will then interact with the environment they were brought up in (and partly chose themselves), which includes the influence of parents, siblings, peers and school. It's clear that neither genes nor environment (nature or nurture) can operate on their own, and the boundaries of both are growing fuzzy. The key message for parents is beware of self-declared experts. There is no magic formula. Do what feels natural and try not to feel too guilty.

Different parenting styles and discipline, bonding and attachment are one thing, regular beatings or sexual abuse of your kids are another, which most of us find hard to imagine. You would suppose that these early life traumas would leave scars impossible to remove. We will next explore how people deal with these experiences, and the epigenetic changes that may explain them.

6

'BAD GENES'

Abusers, criminals and victims

Early on a sunny morning in April, behind the picturesque nineteenth-century church of St Peter's on the 'lawn' near the high street in a small seaside town in East Devon, a small crowd was gathered silently in the graveyard. A few of her relatives and the local clergy looked on nervously as the local police and coroner's office supervised the exhumation of a body buried in a coffin 22 years before. The priest gave a brief blessing as the coffin reached the surface and was placed in the county coroner's van to be taken for DNA tests, before being buried again in a few days' time.

Binny Day was a troubled lady. Her third marriage had just failed and her husband had walked out on her and her three kids. She walked to the cliffs in the quiet seaside town of Budleigh Salterton and jumped. Her twisted body was found a few hours later on the beach 200 feet below by a walker. She left no suicide note. She was only 27 years old and one of twelve siblings. She was pretty and generally well liked in the town.

Only 22 years later did the true reasons for her unhappiness emerge. Her father, William Dance, had served a prison sentence in the 1960s for child sex offences, but was released home after only nine months. Binny had killed herself on the anniversary of his death a few years before.

Her older brother Tommy, who was well known locally, initially for his charity work for childhood leukaemia, was now behind bars for 21 years after admitting raping and abusing multiple young local girls over 30 years. He was only caught after one of his victims watched a similar story on the TV soap *EastEnders* and decided to go to the local police. She said after the trial: 'He turned out to be an evil, manipulative liar. He took away my teenage years, my virginity and also something inside me. It has brought some closure to the last twenty-three years and I hope that he rots in prison.'

The police now suspect that Binny's suicide was the result of being a long-standing victim of sexual abuse and repeated rape by her father and more recently her brother. They are currently investigating if she or any of her children were born out of abuse or incest. However, the story doesn't stop there. Last year her 22-year-old son Carl (who was only five years old when she died) was told by the judge sentencing him that he was 'manipulative, devious and dangerous to young boys'. He admitted 42 offences, including grooming two young boys, and is in jail for a second time after previous assaults and being found with indecent pictures of kids. The police suspect more incidents and accomplices, and Binny's two brothers Maurice and Colin are on trial in Devon for past sex offences. With three generations of sex abuse in a small community the locals are in a state of shock and collective guilt. One neighbour said: 'People in Budleigh now know they are being watched.'

What can this horrific story of three generations of abuse and abusers tell us about ourselves and the human variation in deviant behaviour? Sexual abuse is sadly not as uncommon as you might believe. Studies suggest that depending on how sexual abuse is defined, between 5 and 25 per cent of girls and 5 and 15 per cent of boys will have experienced some form of sexual abuse,[1] usually before the age of 11.[2] On average only about 1 in 10 cases of abuse gets reported to officials, and even professionals fail to report it. Most sexual abuse is actually committed by friends or family and only 10 per cent comes from strangers. Most abusers are men, but in some US studies 5–10 per cent of abusers are women, although these cases are usually milder.[3]

Nor is sexual abuse a modern phenomenon, even though it might appear to be more common today because of the media and greater openness. Records from the Byzantine Empire (AD 324 to 1453) show that despite strict penalties of cutting off noses and capital punishment, abuse was very common. The legal age in Byzantine times for marriage for girls was 12, but often they were married before this and their birthdates altered. There are numerous records of systematic rape and abuse of these girls by their 'husbands' from the age of five in all social classes.[4] The practice is still common and under-reported in many Muslim countries where the victim is likely to be penalised for reporting it.

Although it seems extraordinary, the Budleigh Salterton story supports research in this field showing that abuse does run in families. Studies have shown that at least 25 to 79 per cent of abusers have themselves been abused,[5] so a 1996 US report concluded that: 'while the evidence for child abuse being itself a cause of later adult offending was inconclusive, it was certainly a strong risk factor.' These uncomfortable facts pose a number of salient questions. If we had unluckily been born into one of these families, would we have been abusers ourselves? Did the abusers or victims have any free will? Why didn't the abused women escape or help the other children?

How do we explain three generations of abusers? Is this just a case of 'evil' genes being passed down? Or the direct influence of non-empathetic, violent, abusive parents? Could it just be the result of a harmful environment, bad peer group and education and lack of local social support? Although abuse is generally more common in poorer, less educated families, the latter explanation seems unlikely in this case, as the area is relatively affluent and part of a close-knit community with good schooling. These abusers and victims were not hermits and lived in a cul-de-sac with many neighbours. Studies have not shown any major causal environmental risk factors other than poverty and poor education. Having a poor local environment can clearly make things worse, but overall has only a minor indirect causal role.

So if we can't blame the environment, are genes the main explanation? Most complex behaviour traits are influenced by hundreds or thousands of genes, so as the genes of the parents are mixed every time a baby is produced, if the father's genes for example were mainly to blame this would normally tend to get diluted out in each generation. Yet the grandchild Carl seemed to be just as cruel and evil as his grandfather. There is some evidence that male abusers will select meek female partners who are more likely to be compatible with their personalities, either being easily manipulated towards their ideas or perhaps genetically having a similar lack of feeling for the emotions of others (empathy). If both parents have a lack of empathy, these genes could continue to be passed on and could be one reason why the bad genes don't get diluted quickly across generations. The combination of a sexually aggressive father and a mother lacking in empathy and with a victim mentality can make a deadly parental cocktail.

If genetic susceptibility is to blame, how might this work? The theory deserves some more thought. Twin studies mainly focusing on school have shown a strong general genetic susceptibility (over 60–70 per cent heritability) both to being a bully and also to being a victim, and a proportion of kids (2.5 per cent) are actually both.[6] Some of the susceptibility genes for being a bully, oddly, are also those that predispose to being a victim, not only of physical bullying but also of sexual abuse.[7] However, the idea that victims are in some way genetically pre-programmed is not something we as individuals or as societies like to admit or discuss. This is neither totally surprising nor illogical: personality must underlie these traits, and we have seen that all personality traits have a clear genetic influence. Twin studies have found that being an abuse victim is only weakly heritable, with a stronger effect for physical abuse.[8]

So genes can't explain being a sexual victim, but abuse can affect children – sometimes in quite different ways.

Victims of abuse

Betty was the first-born and larger at birth, but from when she can remember also the quieter, more sensitive and introvert twin. Juliet always took the lead, and was the more extrovert. Betty feels she has always needed her twin more than her twin needed her. Childhood was miserable for both of them – their parents always shouting and fighting, their father often violent towards their mother. They lived in a small council house in Birmingham with an older brother and sister. They were very frightened by their father, who had been in the navy and now worked long hours as a train driver. He, after drinking, would sometimes demand to see them individually. He would sometimes kiss and fondle them and their older sister and would hit them if they ran away.

Betty explained: 'Our mother was cold, antisocial, and never offered sympathy, support or emotion, and other older family members were not much better. She told us she had tried to abort us before we were born. She was often very depressed and had no friends.' Their brother was always ill and often in pain and cried at night, but the parents never got up and the twins had to comfort him. Both their parents came from unhappy homes, having left their homes themselves at 14.

The twins were very bright, and although the parents wanted them to leave school at 14 their teachers persuaded them (by providing generous bursaries) to let their daughters stay on at school. Betty and Juliet both worked incredibly hard and studied together, urging each other on, often till dawn. They realised early on that education and university was their escape route from the family. Both did so well that they were accepted into Oxford University, into the same college, to study Classics. Juliet always did slightly better. After university they both became lawyers in different firms.

Juliet was much more successful and made good money with international companies. She eventually married another lawyer and could afford to have nannies and send their four children to private boarding school, and they had a number of holiday homes around the world. Betty had married a head waiter she met on holiday while still at university and also had four kids, but she gave up work early to look after them. Money is still very tight and she is back in full-time work supporting the family.

Looking back, they have rather different attitudes. Both have suffered depression, but at different times. Juliet now says: 'Although my sister doesn't believe me, I really don't remember any of the bad things in my childhood, it's all a bit of a blur.' Possibly this is a form of amnesia common in child abuse. 'Just after my final exams at Oxford I broke down, I don't remember much but was told I was sobbing uncontrollably for days and had a severe bout of depression. This lasted a year, where I was offered support with counsellors, antidepressants and eventually a friendly GP.' She got back on her feet and into a good job and never looked back.

After an incredibly successful career, Juliet in her early fifties has enough money to retire. She has decided to help others and after a spell working in the charitable sector become a psychotherapist. Although she is reluctant to discuss her past, she realises that she needs to. Betty in contrast was always more open about her experiences and has always suffered emotionally, as well as being generally more susceptible to illnesses than her sister. She has various bowel disorders, as well as the muscular pain syndrome called fibromyalgia, which is often associated with childhood abuse. Betty herself doesn't now think psychotherapy or counselling helps. Although both twins successfully escaped from their family environment and have brought up eight apparently normal children, Juliet still sees Betty as the weak one and Betty sees Juliet as the non-empathetic one. They never discuss their upbringing together, which causes tensions, and as a consequence they now rarely see each other.

Although the twins showed different responses to their abusive childhood, life would probably have worked out much worse for them both if they had left school at 14, got pregnant early and picked abusive non-empathetic husbands as their mother did. Instead, although clearly both affected by their early years, they were protected by their school teachers and the major social and cultural cushion of their university. The story reiterates how even minute differences in emotional reactions to the same trauma between identical clones can lead to large changes in life-courses.[9]

There is a very clear correlation between having abusive parents and later childhood and adult behavioural problems and depression.[10] The common effects of child sexual abuse include depression, post-traumatic stress disorder, anxiety and a propensity to further episodes of victimisation in adulthood. As we have seen before, a correlation doesn't always mean a cause, but there is strong evidence that abuse-exposed twins have a consistently higher risk for psychological disorders than their non-exposed twins.[11]

So do all abused children (and we are talking up to a fifth of the population according to some estimates) have problems in adult life? That was certainly the prevailing view of psychotherapists. However, a summary meta-analysis of 59 US college surveys and seven population-based studies by Drs Rind and Tromovitch from Philadelphia in 1997 emphatically and controversially disagreed.[12] They thought the totally negative and inevitable aspects of damage from abuse had been based on non-representative cases and markedly overestimated in the past. On the basis of the general population and college surveys, they said that child sexual abuse does not always cause harm and reported that 33 per cent of women and 60 per cent of men said they had been unaffected by it. Strangely, a minority of college students actually reported such encounters (presumably fondling by a family member) as positive experiences, and the extent of psychological damage depended on whether or not the child described the encounter as 'consensual'.

This paper generated a wave of protest and showed the professional hazards for researchers in this area. The study was criticised for flawed methodology and selective use of the data, as well as for being 'morally wrong'.[13] The authors responded robustly to the methodological criticisms, but their voices were drowned out. Whether or not it was accurate or morally wrong, the study result was unwanted. The Catholic Church, Dr Laura (the most popular US radio talk host) and the far right anti-homosexual lobby were up in arms, and even the US Congress condemned the study for its conclusions. They claimed it provided ammunition for paedophile organisations to justify their activities.[14] Yet, a recent re-analysis of the data 12 years later showed similar findings to the original.[15] Despite the criticisms and the problems of defining abuse (which confuses legal and scientific definitions), there remains a very wide variety of responses, and a group of victims who are clearly less affected than others.

The different reaction of the twins in our case-study is interesting. Betty never tried to blank out the experience and never had any amnesia or denial, while Juliet, after a brief breakdown early on, decided consciously or subconsciously to forget the past. You could argue that Juliet's avoidance reaction was in retrospect the most successful strategy. Or it could be argued that her early breakdown was helpful to her, although she believed the few counselling sessions she had were unhelpful. This goes against the advice of therapists who urge victims to talk about their experiences early and openly. Cognitive behavioural therapy (CBT) is the mainstay of treatment and has been shown in many studies and meta-analyses to be overall effective compared with no treatment. This established view has been questioned recently. Up to 50 per cent of victims don't respond to CBT treatment and usually give up. Presumably for them, this treatment doesn't work and in some cases may actually be harmful.[16]

Recent studies of veterans from Iraq and Afghanistan, where rates of newly diagnosed post-traumatic stress disorders (PTSD) are very high at over 20 per cent, show that those who never talked about their experiences initially actually did better than those who were openly encouraged to do so. This may relate to the compensation system which in the US pays out $5 billion annually to victims. Clearly this is a disincentive to improvement, but also it is hard to separate PTSD from normal trauma that can improve naturally. Other surveys with stricter criteria show lower rates of PTSD of only around 4 per cent, suggesting a large overestimate of diagnosed victims. These studies are not popular.[17] Authors receive hate mail and death threats. In our modern Western culture sadly it has become very difficult to report openly that estimates of any statistics of disability, obesity, cancer, sexual group or harm of any kind are lower than previously believed.

According to Professor Simon Wessely, a psychiatrist from King's College Hospital, many diagnosed soldiers now report symptoms such as flashbacks, which are actually a new medical phenomenon, never reported in the past. It was unknown after the First World War and some commentators suspect that it could be due to watching TV and Hollywood films which depict flashbacks. Other studies (using a scenario of having been lost in a shopping mall as a toddler) have shown how false traumatic memories of childhood can easily be implanted into many people. The susceptible people then actually refuse to believe it even when told the truth.[18]

This shows just how hard it is to separate real from fabricated subconscious traumatic memories,[19] whatever the cause.[20] It also explains why Juliet's false memories still cause some doubts despite what her sister told her happened.[21] No one is trying to suggest that victims are not genuine or that abuse or the horror of war is wrong. It's just that we need to keep an open mind on perceptions and accept that some people can deal with traumas very well and others cannot, and that some can create false memories more easily than others.

Evil genes?

Grand Rapids, Michigan. A 21-year-old Art and Design student had been meeting friends downtown on a cold windy day in

November when she returned to the deserted car lot to drive home, much later than she had intended. The multi-storey car lot was very quiet on a Monday evening, but she was not alone. As she opened her car door, from nowhere a heavy blow hit her neck, thumped her head against the roof of the car and stunned her. A large powerful hand was put against her mouth to stop her screaming. She struggled and was punched in the face and then pushed over the bonnet of the car and sexually assaulted. An hour later she went to a local hospital and bravely reported the assault to police and underwent a sexual assault examination. The doctors swabbed her vagina and sent the sperm found there to a state crime laboratory. She told police: 'It all happened so fast and it was dark.' She had no time to properly identify her rapist, other than as a large African-American male.

With no fingerprints at the scene and the victim unable to give detectives a description of her attacker, the only chance was a genetic match in the DNA database. Unfortunately no match was found, and her case became yet another unsolved rape. But five years later, the police got lucky. The state crime lab matched the DNA to that of Jerome Cooper, a 36-year-old African-American ex-college track coach, who had been in prison some years later on an unrelated conviction. Like all defendants facing criminal charges in Michigan, Cooper had to submit a DNA sample for the state database. The match seemed to solve the case until they found out he had an identical twin brother, Tyrone, who also had no alibi for that night. Photos of them show that Tyrone is larger and taller than his twin, but otherwise has the same aggressive facial features and thin Mexican-style moustache. They also had other and nastier traits in common. Tyrone and Jerome Cooper both already had a criminal record for sexually assaulting neighbours – a ten-year-old and a twelve-year-old girl.

Their mother Carrie doesn't believe either of her sons raped the woman, though she can't explain the DNA evidence. She told the local paper: 'My sons are not rapists, and they are good boys. I brought them up in church.' She says her sons were falsely convicted previously. 'I've had so many compliments about how respectful my boys are, even from the prison system,' she said. Three of her sons have been convicted of sexual assault, she said. Four of her seven sons are now in prison.

So the Cooper family, like the Devon family, seem to have inherited some bad genes that according to their mother couldn't be tempered sufficiently by their church upbringing, although other studies have shown that religion is usually a beneficial factor, protecting somewhat against crime and alcohol addiction.[22] Carrie Cooper wonders if a relative in years past got away with rape and if God has chosen to punish this generation of her family. 'Things will be visited upon my kids,' she said. 'I think it's a curse upon my kids. I pray to God to lift that. I'm going to leave it in the hands of the Lord.'

The Lord may yet have to have the last word. The DNA sample implicates both twins, but cannot yet identify which one with enough precision to meet the tough standards in the US courts. O. J. Simpson was acquitted of murder – and he didn't even have a twin. *CSI*-style genetic research and technologies are playing an increasingly important role in the field of crime. They are used for two different purposes: first, much as in the context of health and disease, genetic research designs have been used to learn about the genetic factors underlying certain traits, such as aggression, violent behaviour and criminality. In contrast to health research, genetic crime research has been very controversial. Critics oppose what they see as the biologisation of crime; by this they mean the danger of people regarding crime and deviant behaviour as something that is written indelibly into our genes. This genetic determinism of criminality is certainly unfounded, but – as for almost all traits – genetic factors do play a role.

Genetic technologies do however also have an important part in another context, which you will probably know from television series such as *CSI* or *Cold Case*. As most people (besides identical sets of twins or triplets) have unique DNA, genetic profiling has become a key tool in criminal identification. When biological traces are left on crime scenes, investigators can have those traces analysed. The emerging profile shows what particular genetic characteristics the originator of the trace has at a number of places on the genome. When two such profiles are identical (called a match), chances are high that they originate from the same person. Genetic profiles do not, for the moment, tell us anything about a person's genetic disease predisposition or any other traits; they are exclusively used to see whether people have identical genetic make-ups to link them or exclude them from certain crimes.[23]

The police in many countries have problems with twins. Last year I was approached by the New Zealand forensic service. A 16-year-old girl had gone to the police. She said she had met a boy in a bar and had gone home with him. He had turned aggressive and violent and refused to let her leave and then hit her repeatedly and in a cruel and calculated way raped her. The police traced the boy from her description. The boy, who was just 18, was arrested. It seemed a straightforward case – until he said he had an identical twin brother who was also in the bar that night. Both twins said they had got drunk and couldn't remember what happened. The police didn't believe their story, as they both had records of violence and petty crime, and wanted to prove this forensically. They had collected a semen sample after the police examination.

The DNA that is easily extracted from the semen and then matched to the blood of the suspect is usually enough evidence to convict a rapist even without identification. Even in identical twins, fingerprints which are very similar are not completely identical and can be distinguished by experts in court. However the DNA from sperm, just like every other cell in the body, carries the exact same classic DNA signature (sequence) in one identical twin as another. The police used current forensic methods which haven't changed for 25 years and compared the two samples, using 13 highly variable genetic markers (called the CODIS system), with the national database. The results were completely identical. They couldn't prove which twin the sample came from and were going to drop the case. I explained that they could separate them if they really wanted to. They could find subtle epigenetic methylation differences in the DNA, or possibly they might differ in some rare gene mutations. We are now working on identifying these differences. Unfortunately both these methods are still considered experimental, and would get ripped apart in court. So male identical twins with criminal predispositions are potentially getting away with it – for the moment at least.

A criminal – or a victim of genes?

On being caught red-handed, many criminals plead insanity, but what about blaming your genes for diminished responsibility? Would that be a legitimate defence? It may sound crazy but a few have tried it. The last to do so was Tony Mobley.

His victim was kneeling and begging for his life when he pulled the trigger. He was convicted for cold-bloodedly murdering 24-year-old John Collins with a shot to the back of the head. His victim was a quiet man who had the misfortune to have been managing a Domino's Pizza store at the wrong time. Molby had been on a crime spree of at least six armed robberies in the preceding weeks. He showed no remorse and told his guard: 'If that fat son-of-a-bitch had not started crying, I would never have shot him.' He jokingly told another guard that he was 'going to apply for the night manager's job at Domino's because he knew they needed one'.

After several trials and mistrials the 26-year-old Mobley was eventually sentenced to execution and kept on death row in Jackson, Georgia for 13 years. Mobley had a very troubled background and has been diagnosed with post-traumatic stress disorder, paranoia, other anxiety disorders and organic brain damage. He went through ten different residential placements. He had been out of control since the age of 11, and a series of special schools and psychologists had not altered him.

Tony Mobley did not come from great family stock. His grandfather was violent and abusive. His great-uncle went to jail for murder, one cousin beat up his wife with a gun, and another let his friends rape his two daughters. His great-grandmother was also violent and beat up her daughter-in-law. His eventual attorney argued against the death sentence on the basis of mitigating circumstances, namely that he probably carried a genetic mutation in his serotonin gene or an extra Y chromosome. Both defects had been associated in genetic studies with aggression, violence and criminality.[24] He wanted the court to allow this to be tested for and used as evidence. The Supreme Court did not agree. He was killed by the state of Georgia by lethal injection with an audience of onlookers, family, victims and journalists behind a glass screen in 2005.

This example shows that genes, while sometimes an explanation of causal factors, are not a good defence in court. They are seen as immutable and so in contrast to the commonly used mitigating circumstance of a broken home and absent father the defence is rarely used.

'There is no legal defence to his crime,' said Daniel Summer, Mobley's attorney at the time. 'There is only the mitigating factor of his family history. His actions may not have been a product of totally free will.' Murder, rape, robbery, suicide, 'you name it', the Mobley family has had it, he said. Sumner used the deadly combination of the wrong genes plus the wrong environment to explain Mobley's psychopathic behaviour.

Simon Baron-Cohen, an expert in autism at Cambridge (and cousin to Sasha, the inventor of Borat), has recently advanced his theory that rather than regarding these male offenders as evil psychopaths with vaguely defined personality disorders, we should label them as having zero empathy.[25] These are people on the extreme end of normality who cannot relate to their victims or feel remorse. Even the Austrian Josef Fritzl, who imprisoned and raped his daughter for years, apparently had little idea of the harm he was doing until he listened to ten hours of her testimony at this trial.

There is good evidence that empathy – the ability to put yourself in someone else's shoes – although a typical human trait, does have a wide range in humans. These empathy differences between us show moderate genetic influences of 30–70 per cent. Up to 70 per cent of borderline personality disorder cases, who are often very low in empathy scores themselves, report having been sexually abused.[26] A high percentage of child abusers and adult rapists also show low levels of empathy,[27] so there is a close overlap between lack of empathy and being both a victim and perpetrator of abuse.[28]

So what are the overall genetic influences on criminality? Many twin, adoption and family studies have shown a strong heritable component. One study of 16,000 US military twins showed a 15 per cent concordance in dishonourable discharge for identical twins compared with only 2 per cent in non-identicals.[29] Other large Swedish studies of 1,400 pairs and their parents show heritabilities for aggressive behaviour and psychopathic personality of over 60 per cent.[30] Studies of several thousand Vietnam-era twin recruits showed that previous childhood antisocial behaviour was strongly influenced both by genes but also by family environment, whose effects largely disappeared when they were older.[31] In Denmark, where they can legally track twins and their criminal records centrally, the same pattern of genes and diminishing family influence was seen.[32] The largest estimate for family influence on childhood antisocial behaviour was 16 per cent.[33]

Of course we have not yet looked at the greatest known genetic risk factor for criminality – one that carries a 15–20-fold risk. There is a 50-50 chance that you will be a carrier. It is called the Y chromosome and determines if you will be a male.

The unique Dunedin Study was ideal for looking at how genes and environment interact in male criminality. They ignored females for this study, as their risk was too low, and followed over 400 males over 26 years in New Zealand to look at predictors of criminal behaviour. The focus of their research was a gene that creates an enzyme called monoamine oxidase-A (MAOA), which 'mops up' extra amounts of neurotransmitters in the brain that are known to influence aggression. Boys who also suffered from physical, psychological or sexual abuse as children were nine times more likely to engage in criminal activity if they also had low levels of the MAOA gene than if they had a normal amount. The presence of the gene variant on its own had only a modest effect (increases of 50 per cent) on children in normal non-abusive families.

Other studies since confirm these results, albeit with less dramatic effects. Most also show that alcohol abuse (which is also partly genetically controlled) is another major risk factor.[34] So a combination of susceptible genes and an abusive childhood is a recipe for disaster, and to date only a handful of genes – a fraction of those likely to be involved – have been explored in this way.

All these studies of antisocial behaviour, criminality or empathy disorders, while confirming a strong genetic influence, also strangely show that the agreement between identical twins is usually only about 35 per cent. This means that over half to two-thirds of identical twins, despite being in the same environment, don't become psychopaths, violent criminals or rapists like their brothers. While some twins may have alone experienced a trauma such as being abused that their co-twin did not, in many pairs no obvious difference in major trauma can be seen.

Brains, psychopaths and victims

What is going on inside the brains of these people? Clear MRI brain-scan changes can be seen in abuse victims as well as psychopaths and those lacking empathy. Brain regions most affected are all in the empathy circuit such as the amygdala and the prefrontal cortex.[35] In a recent study from the Institute of Psychiatry in London, adolescent twins, some with early psychopathic traits, were scanned.[36] The researchers found that the more grey matter you had in empathy-associated regions the more likely you were to have antisocial or psychopathic traits. Excess grey matter was moderately heritable and is believed to be a sign of brain immaturity, and those affected twins with more grey matter were in essence empathetically more immature than their co-twins. But what was the mechanism for these changes? How could genes such as those influencing empathy, which provide the susceptibility in both twins, have been changed? Could epigenetics be responsible in the twins, or even in the parents? The animal studies again provide clues.

The Montreal rodent studies we discussed in the last chapter show us that maltreatment in the first week of life causes certain genes to be switched off epigenetically.[37] The glucocorticoid receptor gene controlling the stress response is the best example, where a methyl group is attached and so the gene cannot be expressed. This leads to a cascade of changes in many other genes related to emotion and stress, and can last a lifetime.

It was thought that any epigenetic marks during life would get completely erased to prevent them being passed on to offspring. However this is not the case. This methylated gene will be present in the sperm or egg of the child, and then usually passed on to the next generation when they reproduce. This means that maltreated pups (or humans), when they become parents, because of their abnormal methylation often fail to bond with their own children because their empathy or bonding genes are not working normally. This leads to an increasing cycle of dysfunctional parenting or abuse. As well as passing on the bad traumatic experience itself, the long-term genetic susceptibility is there in the families, plus the shorter-term epigenetic changes. This combination produces a nasty cocktail leading to potential multiple generations of antisocial behaviour or abuse.

Recent studies in humans have backed up the rodent experiments and shown that at the extreme end, suicide victims who had reported abuse were more likely to have epigenetic changes in crucial hippocampus areas of their brains compared with non-abused suicides.[38] It appears that as well as the glucocorticoid receptor gene controlling stress, the genes controlling dopamine, serotonin and the cuddle hormone – oxytocin – levels in the brain are being switched off epigenetically in these families[39] and passed on from one generation to the next – if they survive long enough to reproduce.[40]

Bruce Perry has pioneered the use of MRI scans in badly abused or neglected children. He and others believe that the brain is like an underused muscle that can, with emotional support and mental stimulation, regain much of its potential (if not left too long). Similar brain scans can now accurately diagnose a paedophile.[41] If he is correct, it might also be possible for criminals and abusers to regain some empathy.

So should we be less fatalistic about families like Binny's, the Coopers or the Mobleys? The traditional view is that if genes are largely responsible nothing much can be done. However, although antisocial or criminal behaviour is heritable and more likely to also occur in twins or siblings of a criminal, it certainly isn't inevitable.[42] In fact the majority of co-twins are socially normal and would be wrongly labelled as likely criminals. These provide hope. The animal and some early human studies show us that these post-traumatic epigenetic changes in the brain are in part reversible.[43] This can be achieved by early fostering with a cuddly mum or by using conventional drugs that block the hormones oxytocin or dopamine or certain epigenetic drugs called HDAC inhibitors (like TSA) or methyl donors – drugs or foods that can deliver methyl groups and change the gene function.[44]

Failing families

The widely held view that 'genes = fatalism' has since the war so pervaded our minds and precluded public debate that it has become practically a taboo topic. Even comprehensive reviews by recent official and inter-governmental groups,[45] or other influential department of health reports advising that intervention programmes focus on social and economic factors,[46] strangely don't mention the 'G' word even once in hundreds of pages. Despite the evident contradictions of optimal parental advice and the current evidence that genes are the main known cause of childhood problems, there is a major political drive to intervene. The strategy is to provide parental advice in high-risk failing families to prevent dysfunctional and antisocial kids. Intervention programmes like the Nurse Family Partnership (NFP) – the most rigorously tested programme of its kind – are making a huge political impact in the US and around the world.

David Olds, a paediatrician in Colorado, started this charitable programme in the 1970s targeting young (often unmarried) poor mothers at high risk. Trained nurses would visit mothers at least twice a month and observe and offer advice. They then conducted randomised controlled trials comparing them with normal services over the next 15 years in Elmira, New York, and in Memphis and Denver to see the effect. The results were spectacular. NFP-trained mothers had fewer unintended pregnancies. They were four times less likely to abuse or neglect their children, or to misuse alcohol or drugs, and were less likely to need welfare support or be unemployed. The nurse-visited children had 50 per cent lower arrests, 80 per cent fewer convictions, significantly lower use of drugs, alcohol and tobacco, and less promiscuous sexual activity, than the control group. The benefits extended for up to 15 years after the programme ended.[47] President Obama endorsed the programme and recently committed $8.6 billion over ten years to the development and refinement of the Federal programme across the USA from 2010.

Many other programmes in other countries are being commissioned, costing millions now with the hope of eventually saving billions in 20 years' time when the children have grown up. Independent reviews and government reports of all the different intervention studies worldwide have been generally supportive.[48] They all show they are most cost-effective for the high-risk groups. However, the experts examining the studies still have some problems. They cannot yet say which type of intervention works best, or at what time point or duration, suggesting that the exact mechanism is still unclear.[49]

What is clear is that if it is the parenting skills that are being changed directly, this contradicts the evidence from the genetic studies showing little influence of parenting. Remember that the studies showed less than 5 per cent reversible environmental influence for personality, and at the most 16 per cent for antisocial behaviour. Some of the scheme's success undoubtedly comes from improving the social and economic factors of these mothers and their ability to overcome these problems.[50]

But the experts also believe that changing the understanding of the parent towards the individual child's needs (and presumably their genes) can help. Perhaps the key is altering and reversing the destructive one-to-one bond between mother and child. When this bond or interaction goes wrong it leads to dangerous alterations in gene methylation that modify their genes affecting emotion and behaviour. Thus, intervening by reducing maternal stress or other simple measures (like epigenetic drugs) may alter or soften the epigenetic signals that might otherwise remain permanent, preventing transmission to the next generation and continuing the vicious cycle.

A new theory recently summarised and tested in US adolescents by Jay Belsky and Kevin Beaver has suggested that genes could be acting in ways we hadn't anticipated. They found that some children are just genetically programmed to react in an extreme way (either good or bad reactors) depending on circumstance. Others, by contrast, can be totally resistant to change whatever the positive or negative environment. Kids with certain forms of the (usual suspect) neurotransmitter genes influencing dopamine and serotonin were more likely to react strongly in one direction or another than those with other forms.[51]

Belsky and Beaver proposed reconceptualising such genes as 'plasticity genes' rather than vulnerability genes. This means that if the worst kids that responded most negatively and became delinquents are then put in an ideal environment, they might actually rise above passive kids in terms of social skills and IQ. This is clearly a big hope of the intervention programmes, but we will have to wait and see if reality matches the idea.

Epigenetic signals are hard to detect in traditional genetic and epidemiological studies, which explains why it has taken us so long to discover them. Why we have evolved to have these epigenetic mechanisms make us so unpredictable is intriguing, but our development as infants and children points clearly to prioritising individuality and being different to our brothers and sisters. This phenomenon might also explain why twin and adoption studies show negligible effects of common environments on kids, because each individual (even identical twins) can react so differently to them.

Plasticity explains why coming up with any perfect parenting plan is unlikely to work for all kids, and it is proving hard to pin down the best interventions for failing families. It also explains why some kids are relatively immune to their dreadful experiences. Nevertheless, putting society's resources into trying to alter the bad behaviour and therefore modify the genes of the small group of the most high-risk families could just produce the dramatic results that we need.

7
'THE MORTALITY GENE'

Hearts, famines and grandparents

When my father went to bed one night complaining of indigestion and failed to wake up the next morning it was a shock. The subsequent inquest and post mortem showed he died of a heart attack. He was only 57 years old and had no obvious risk factors except a lifelong dislike of exercise. He didn't smoke, had a normal diet and had no previous heart problems; both his parents had lived twenty years longer than him. The underlying reason for his premature death was a mystery. He was a medical pathologist, an expert on heart disease and cot death syndrome. At the time of his death in 1981 he had been writing the outline draft of a book entitled *Social Pathology: The Social and Psychological Causes of Disease*. Executive stress was the trendy subject of the moment, and the role of genes in causing death and disease was barely discussed.

When twins die within minutes of each other, as does occasionally happen, it is the classic type of news story that makes headlines around the world. Such stories help to reinforce the traditional view that you can't escape your genes - and the timing of natural death is pretty much pre-ordained. But the vast majority of identical twins do not die within hours of each other, and one usually survives the other by several years. One of the world's largest twin studies followed up 40,000 Swedish twins for 36 years. In the pairs where one died of heart disease, only 40 per cent of male and 30 per cent of female identical twins also died the same way.[1] In the cases of non-fatal heart attacks, the correlation between identical twins was even weaker.

Despite sharing the same DNA and genetic make-up as well as similar environments, why is it that most identical or non-identical twins, let alone siblings, do not usually suffer from the exact same diseases? The structural DNA in every one of the trillions of cells in our body is, for practical purposes, identical, as are the individual genes and gene variants. As most cells in our body replicate and divide to keep us growing or alive, this genetic material is passed on exactly to the next generation of cells. In this way our cells can be thought of as mini-replicas of ourselves. Now DNA makes proteins, themselves made up of amino acids, which are the main drivers of all the chemical processes in the body. There are four times as many proteins as genes, and these proteins can also activate genes themselves. In other words, it is not the case that every gene makes one protein, but some genes make several proteins and some proteins are produced by the joint work of several genes. The possible proteins combinations are endless and the same genes can generate very different messages in the cells.

The cells that make up the 200 different tissues in our body (by which I mean heart, lung, bone, skin and so on) all behave differently because they have different proteins which vary in each cell type with different functions. How can this be? Clearly, although every cell has exactly the same DNA and the full complement of 25,000 genes, they are not all used or expressed equally. It is this variable gene expression that distinguishes the cells from each other. This is caused by what are called epigenetic signals like methylation (adding a methyl group) which we have discussed before, or the more complex histone modification (chromosome folding), which controls the expression of the genes. These messages get passed down the generational line each time a cell divides, making sure that a heart cell as it grows or ages, importantly keeps behaving like a heart cell and doesn't get distracted and revert to some other kind of cell.

We believe that hundreds if not thousands of genes are involved in most complex traits like heart disease, each with a tiny individual effect.[2] To use a musical analogy, the genes are like the 25,000 pipes of an organ used in some concert halls (our cells), whereby hundreds function in harmony, being opened and closed by the mysterious (epigenetic) organist to control the music. It no longer makes any sense to talk of a 'heart attack gene' any more than you can say there is a single note for Bach's Fugue in D minor.

Because multiple genes work together their effects are hard to predict and minor subtle changes in a few genes can have major consequences. This is where epigenetics and methylation feature. Small changes in a risk factor like smoking or gaining weight can alter methylation in a large number of genes and so affect their function in many ways. This is why if we want to better predict clinical outcomes like heart attacks we may need to look elsewhere.

Fate, feast and famines

At 4 a.m. on a cold winter's morning close to the Scottish border a strong wind was battering the windows of Andre's house, but it was the strange aching down his neck and left arm that woke him. He hadn't done anything out of the ordinary the previous night, just sat in the pub as usual, had a few drinks and his weekly Friday night curry. When the pain kicked in again, harder and vice-like, he suddenly felt like retching and prodded his sleeping wife. She woke to find that he seemed to have stopped breathing.

At the age of 67, Andre had just suffered a heart attack; his doctor would later note in her files that he had suffered a myocardial infarction due to blockage of his coronary arteries. Like two thirds of victims nowadays he survived this first attack, giving him a chance to re-evaluate his life. He would describe himself as well-built, at nearly six foot tall, but with a bit of a weight problem that had afflicted him since childhood, despite his occasional attempts at cycling. His older brother, who had moved to the US, and his sister, who still lived in the Netherlands, where he had been brought up, were both fitter and slimmer than him and had no heart problems.

Until the twentieth century heart disease was unusual. In the West it is now the commonest form of death. In countries such as China and India, where death from heart problems was virtually unknown in rural areas until the last thirty years, the situation is changing rapidly. 60 per cent of all worldwide deaths are now due to so-called Western illnesses, such as heart disease and diabetes. Genes cannot explain the massive increase in the rates of heart disease over the last seventy years, nor can it be down to slow Darwinian selection. Though identified lifestyle risk factors such as obesity, poor diet and smoking, have increased.

So Andre's problems could just be a result of his unhealthy lifestyle. Although trends are improving, Scotland, where he lives, has high levels of heart disease and instances of poor diet. But let's not blame the Scots alone. Could we also blame the Germans? Andre's destiny could well have been altered before he was born when his Dutch mother was pregnant.

In the winter of 1944 the forces of Nazi Germany were in controlled retreat after the Normandy landings in the West and the Soviet offensive in the East. Attempts by the Dutch population to help the British advance from Arnhem with attacks by partisans and a National railway strike had failed. The Germans took their revenge. They stopped food reaching the west of the country and flooded huge areas of farmland at the same time as an unusually cold winter hit transport and natural food supplies.

Roughly half the Dutch population starved, living on less than 1000 calories a day. Over 20,000 did not survive, those that did, like Andre's mother, had to beg for food and resorted to eating pets, rodents and tulips to stay alive. After the war a group of British and American doctors were flown in and saw the opportunity to look at the effects of starvation on long-term health, in what became the Dutch Famine Birth Cohort Study. In the rubble of devastated cities like Rotterdam, they managed to find key obstetric records, which showed the markedly lower birth weights of babies exposed to maternal malnutrition in the crucial last few months of pregnancy and the paradoxically larger birth weights of babies exposed in the first three months.

Following the lives of these children, the researchers noticed that when the boys born in the famine had reached 18 years of age and enlisted in the army they were considerably fatter than their friends born in the east of the country or even neighbours or relatives born just six months earlier or later, who had not suffered from prolonged fetal malnutrition. As they grew older in their 1950s and 1960s they found that rates of obesity, heart disease, diabetes and schizophrenia were also greater. Similar research in the 1980s found that 5,000 British men born in Hertfordshire who had the lowest birth weights, also had the highest risks of heart disease 60 years later.[3] So some of Andre's current propensity to obesity and heart disease could well be linked to his poor nutrition as a fetus. The studies are still ongoing into whether the risk of famine extends to the third generation.

Other such tragic 'natural experiments', along with good historical detective work, could help support these findings. Hidden from the West for many years, the Chinese famines during the so-called Great Leap Forward affected large parts of many rural provinces. Like in Stalin's Russia twenty years before, this was caused by collectivisation policies and the pseudo-science of Lysenko's crazy agricultural ideas. Up to 50 million Chinese peasants are believed to have died.[4] Mao used the reforms to also pay off massive debts to Russia in the way of grain, which Russia ironically needed because of its farming failures due to his own collectivisation policies. A third of all the grain from villages was requisitioned for this purpose and Mao was quoted as saying, 'When there is not enough to eat, people starve to death. It is better to let half of the people die so that the other half can eat their fill.' Fifty years later, the official Chinese position on the famine remains a sensitive issue.

Although only very basic official records are available recording births and deaths from the provinces, there are huge numbers of them. They show that starving women had massively reduced fertility rates in the following year. In regions with sudden drops in fertility, researchers surmised that famine conditions were present for at least 12 months before. They then followed up the rates of psychiatric diseases in these women's children twenty or more years later compared to children from non-famine areas. Children born from starving areas showed twice the normal rates of schizophrenia as adults. The same pattern was seen in the Netherlands, and in the Anhui and Guangxi provinces in China, regions 1,000 miles apart - showing the risk was worst in rural areas where the famine hit hardest. Epigenetics could be the mechanism for both these remarkable observations in China and the Netherlands.

It is possible, however, that the consequences of both the

Chinese and Dutch famines were simply a direct effect on the fetus due to lack of nutrition. Knowing whether such risks could be passed on to the third generation would therefore be crucial to determining whether epigenetics played a role. Although not many countries keep detailed historical records of harvests, obstetric and nutritional details as well as causes of death across generations, luckily for us, the Swedes have long been keen record keepers.

The small farming community of Norrbotten in the north of Sweden was so isolated and self-sufficient in the nineteenth century that if the harvest was bad, as it was in 1800, 1809, 1812, 1828 and 1856, people starved. If it was good, they gorged themselves excessively. One study focussing on the parish of Överkalix and a cohort of 330 adults born in 1905, estimated the food availability for their parents and grandparents on the basis of harvests and food prices over the period from 1803 to 1849. The oocytes (eggs) in the girls and sperm cells in boys (known as germ cells) are formed differently to normal cells with half of the complement of chromosomes and stored separately to the rest of the body, eventually in the gonads. The study focussed on the period between the ages of 9 and 12 years, just before puberty, when the DNA in sperm, which is usually well-protected in the scrotum, is moving downwards and probably most susceptible to epigenetic modification.

The first results were a surprise. They showed that adults whose grandparents had gorged themselves in feast years, died on average six years earlier than those whose grandparents survived famine years. When they expanded the group and looked at their genders, the findings became even clearer. The paternal grandsons of men exposed to famine before the age of 12 did rather well and were actually less likely to die of heart attacks.[5] Over-eating grandfathers seem to have passed on to their grandchildren not only an elevated risk of earlier cardiac death but also a four-fold risk of diabetes.[6] The findings were most clearly seen in males but a similar trend was found in women. Yet again the trend was found only within genders: women were affected only by the habits of their grandmothers, men only by their grandfathers. So something happened to the eggs or sperms of the respective grandfathers and grandmothers during the famine which affected their grandchildren two generations later. A follow-up study in Bristol showed that early childhood smoking (before 11 years) in 166 fathers produced fatter sons, but not daughters.[7] Over-eating before puberty had the same toxic effect on the next generation.

The only other similar study in humans comes from the study of Betel nuts. These are commonly chewed around the world by 600 million people in Asia, and Taiwan in particular. Taiwanese researchers already knew that its regular consumption was linked to increased rates of cancer of the mouth, but now found it was also related to increased risk of later obesity and diabetes in men and importantly their sons.[8] This result suggested that even exposure after puberty could still cause epigenetic changes in the sperm when it had reached the relative safety of the testicles. This may be because men continue to produce sperm all their lives and sperm DNA can be influenced over many years. Similar experiments in male mice fed betel nuts produced the same results: offspring with increased rates of diabetes, particularly males.[9]

The fact that changes in weight and diabetes risk occurred through the male line via sperm, rules out a direct effect of fetal nutrition as the cause. The environmental signals or changes to the genes must have somehow been passed on genetically to the grandchildren. One of these key genes that researchers examined is involved in the growth of the fetus (IGF2). This was abnormally methylated in sons and daughters of Dutch famine victims tested 50 years later when compared to their siblings born shortly before or after.[10] Other studies of Dutch famine babies suggest they have a significantly lower risk of colon cancer, one of the commonest cancers in adults.[11] They found the important promoter area of the cancer enhancing gene was over-methylated, effectively turning it off.

These historical studies tell us is that what your parents or grand-parents were doing or consuming before you were conceived can have a major influence on your life at least two generations later. What we thought might be an adverse environment like starvation, while leading to a high schizophrenia risk, possibly through the stress pathways we discussed earlier, could strangely also confer some advantages in protecting against cancer. By the same rules, over-eating at different times can also have good and bad consequences in future generations. Separating cause and effect is not easy when studying socio-economic factors in health and disease, but twin studies can again help us tease them apart.

Stress, social circles and simians

Joyce and Margaret were identical twins born in post-war Dulwich, in south London. Their family moved to Reading, 30 miles away, to a very average middle-class, three-bedroom semi-detached house, where they grew up with another sister. Their father worked for the post office and their mother was a part-time nursery teacher. They had a fairly happy childhood, didn't get into trouble and had no major illnesses that they remember. They looked alike and shared most things, until they left home at 18. They both first went to secretarial college for a year, but Margaret got bored and found a job at Selfridges, while Joyce completed her training and got a job as a junior secretary in a law firm. They started to move om different social circles. Margaret was the first to announce she was getting married – to Frank, a friendly plumber from Walthamstow. Joyce followed the next year and married Anthony, one of the junior lawyers in her firm.

They had five children between them and led generally happy lives, but their financial statuses started to diverge when Anthony was head-hunted by a big city law firm and unexpectedly early became a partner. Frank had some recurrent back problems and several times had to take 6 months off from his own plumbing business, which suffered badly. Margaret went back to work to keep the family going, and thoughts of buying their own house were put on hold, while Anthony and Joyce moved into a large family house. Although less than 5 miles apart, they lived in different social worlds, causing a strain on their relationship.

When we saw the pair for a twin examination, 35 years since their paths had diverged, some of these differences were now evident. Both twins had smoked continually in their twenties, but Joyce had managed to quit after getting married, while Margaret admitted to a 15-a-day habit until she was 40. Both still liked a drink: Margaret preferred port and brandy, Joyce white wine. Both were the same height, 5'4", but Margaret was about 15lbs (7kg) plumper than Joyce, who went to exercise classes. Margaret, despite her plumper cheeks, looked older than her sister, who also had more of her teeth. When we tested their DNA we found that Margaret had a shorter average telomere length (by 200 base-pairs) in her white blood cells than Joyce. This suggested that although they had the same chronological age to within a few minutes, the biological age difference between them now reflecting the age of their cells was around 7 years, even accounting for their smoking.

We found the same social effects as in Joyce and Margaret in a formal study of several thousand British twins, which allowed us to remove the effect of genes by comparing within identical pairs. Social status was clearly correlated to a biological marker of early ageing, or telomere length. [12]

A separate study of 10,000 British civil servants showed a three-fold increase between those at the top and bottom of the administrative tree in terms of risk of heart disease.[13] However there were, as expected in the lower-level messengers and cleaning staff, many more bad habits, like smoking, poor diet, and lack of exercise. Yet even when all known risk factors were combined and accounted for, they only explained one third of the excess mortality: the majority remained unexplained. The authors of this study believed the missing factor could simply be stress. This in turn may be due to a perceived feeling of loss of control caused by the relative position of an individual in a society – something unaffected by absolute wealth. Epigenetics could explain how stress translates into health problems. It is well known that the lower in the perceived social pecking order someone is the more long-term stress they are under. This is true in rodents, primates and humans.[14] A recent study in monkeys placed females in different social environments and groups and noted their level of stress and health. The worse their social position in the new group they joined, the worse their stress levels and also their immune system. The mechanism at work appears to be methylation, epigenetically deactivating their immune genes.[15]

The telomeres we measured in the DNA of the twins are a marker of the accumulated stress in a cell: they act as protective caps (like plastic on boot-lasses) on the ends of the chromosomes to stop them being eroded. As you age or get cellular stress due to smoking, obesity, chronic illnesses or social status, these telomeres get progressively shorter.[16] Once they shrink below a certain number of base-pairs (around 13), they no longer protect the chromosome and the cell auto-destructs. The more the relative social differences between groups, the more the cells are subject to stress (often called oxidative stress) and the more they lose telomeres.[17] This increases the risk of early heart problems and many other common age-related diseases.

The male residents of central Washington can expect to live 17 years less than their neighbours in suburban Maryland, and there are up to 25-year differences between residents of some areas of Glasgow.[18] These mortality differences cannot be explained by the classic risk factors of nutrition and lifestyle alone. The poor groups in Maryland and Glasgow are under more stress and have less control over their lives. They have the same low life expectancy as the average male in rural India. Chronically stressed cells and their owners do not grow and replicate to the same degree. Although the human studies haven't yet been performed, animal studies suggest it is likely that chronic stress epigenetically switches off the genes that are important for normal development and the heart. The same epigenetic stress message will continue to be transmitted through the next few generations, perpetuating the tendency – and possibly the social divide.

The key message here is that the environment and exploits of our parents and grandparents influence us in a number of ways – modifying our growth, altering our brain development and affecting our risk of diabetes and heart disease. These environmental stresses have been passed to us epigenetically, by so-called 'soft inheritance'. The reason that lifestyle and environmental risk factors for disease and our mortality have been so hard to pin down may be that we have been looking in the wrong place and wrong time. We should have been sending out questionnaires and surveys to our grandparents 100 years ago.

The bigger question has to be how this affects the future. What we eat today will have a major impact on future generations, regardless of what they eat. The answer as to how they will fare can be found in our present environments and diets – something we will explore in the next chapter.

8
'THE FAT GENE'

Diet, worms and wine

My first practical encounter with real obesity was as a junior doctor in the 1980s on a 12-month exchange programme in Brussels. I was summoned by a nurse to a room in the smart newly opened and novel six-bed obesity unit of St Luc hospital to help move a patient. Four other nurses were already there, looking red-faced and stressed. Between them on the bed was the largest woman I had ever seen or even imagined. She was 45 years old and weighed 220 kg. Although she was naked, rolls of fat covered her body and modesty like curtains. She had been brought in the day before by the local fire brigade, after collapsing at home. The local ambulance crew had been unable to lift her from the floor. It took five of us just to roll her onto her side.

She stayed in hospital for eight months, until she was fit and slim enough to walk a few paces, after losing 60 kg. Her recovery wasn't helped by her diminutive and skinny husband, who we later found was smuggling in biscuits to supplement her 800-calorie diet. He liked her very big and fat and may well have had a bizarre sexual perversion called 'feederism', by which men get sexual pleasure from helping keep women very large, a quirk that was unrecognised at the time.[1] Since my early experience 30 years ago, obesity clinics have opened up in hospitals around the world, and it has now become a recognised medical speciality as the epidemic has taken hold.

In the 1970s medical textbooks didn't even mention obesity;

now 30 per cent of Americans and around 20 per cent of Britons are obese, with a body-mass index (BMI) of over 30. BMI is the way obesity is compared and classified across the world and is calculated by dividing your weight in kilograms by your height in metres squared – e.g. 100 kg/(2 m × 2 m) = 25. The majority of the population of both countries are now classified as overweight (a BMI of over 25). Most of us aged over 40 remember being at school with maybe one overweight or fat kid in the class. Now it is commonplace, with one in four kids already overweight or obese by 12 years old. The average 55-year-old US white male has since 1974 increased in weight by 10 kg (22 lb) and the US female gained 11 kg (24 lb).[2] In Florida, which is currently mid-table in the US fat league tables, 63 per cent of adults are overweight and the associated disease diabetes, which causes early heart disease and death, occurs in an amazing one in ten people. Another 12 common diseases are increased because of obesity.

The rest of the developed world is quickly following America's cultural lead, and an unlikely country like Greece is the fastest-growing in Europe, despite eating a so-called 'healthy' Mediterranean diet. This recent change across the world, leading to nearly 2 billion overweight individuals globally, has been caused by a recent massive over-abundance and access to particular types of cheap food – particularly refined carbohydrates and fats – unavailable to our ancestors. This is causing an epidemic of overnutrition that we are only just beginning to come to terms with. But is this sudden change in weight just a result of overeating, or could genes that take hundreds of generations to change still be responsible?

A combination of twin and family studies have shown us that obesity, whether measured by total weight, body-mass index or fat levels measured by scans, is strongly heritable: most estimates range between 60 and 70 per cent. Most identical twin pairs are within a few kilograms of each other's weight, even in old age. The world's heaviest twins according to the *Guinness Book of World Records* were the McCary twins (who used the exotic stage name McGuire) from North Carolina, who started life as wrestlers, then performed carnival motorcycle stunts and weighed in at roughly 730 lb (330 kg) each. They had increasing trouble jumping over buses though, and Billy died age 31 in a bike crash. Benny, amazingly, lived for a further 21 years until he finally succumbed to heart disease.

So far over 30 definite genes regulating fat have been discovered by worldwide consortia we work with. The numbers needed have been huge – 250,000 subjects in the last study. At first the media got very excited about predictive testing in infancy, until the tiny size of the individual effects of these genes was realised.[3] On average individually they raised your obesity risk only by about 5–10 per cent, and even when combined together, they can't explain or predict accurately who will get the disease, as they only explain a few per cent of the total variation in weight between people. But the study has told us a lot more about the biology of obesity.

Before the study we believed that different metabolic rates and different types of fat were the key genetic factors in why people differed. We now know that the brain may be more important. The first and strongest gene found so far is called FTO, and is expressed in the brain, especially in the key reward centre of the hypothalamus in the base of the brain (hypothalamus means under-chamber in Greek). For some rare humans who have two copies of the variant, chances of being obese increase by up to 70 per cent. Experiments in rats and observations in humans have shown that having different variants of FTO genes directly alter the chosen diet – influencing total calories and fat content of food eaten and may release oxytocin

– the cuddle hormone.[4] Other forms of a newly discovered gene called amylase we helped uncover dramatically alter the wish for starch and fatty foods and influence obesity.[5]

Most of the other 30 or more genes associated with obesity found to date are actually expressed mainly in the brain, not the fat or intestines or liver, where metabolism mainly occurs. But all is not lost: even if you have more than your share of the dodgy genes you can reduce their influence by at least 30 per cent just by being physically active, which switches off (epigenetically) the effects of genes like FTO.[6] So overeating seems to be primarily a brain problem, either of greed through the reward centres of the hypothalamus, or not feeling full quickly. But what makes us or our genes decide how much to put into our mouths?

About ten years ago several novel hormones were discovered which help regulate when we feel full, but so far only one hormone, Ghrelin, has been found that controls when we feel hungry. We know there is a gene that codes for the Ghrelin protein, and so the threshold and intensity for the hunger signal has a general genetic influence. While normal twins would be expected to have similar Ghrelin levels, a study of identical twins has shown that the hormone levels differ significantly between fat and thin discordant pairs, suggesting that once an individual has gained fat, this new environment and possibly diet alters the expression of Ghrelin.[7] Alternatively, differences could have occurred earlier in life, although the researchers didn't have childhood samples to test.

This controlling influence on Ghrelin probably works via epigenetic mechanisms, and the study suggested that some people are 'genetically' more sensitive to positive or negative changes in their diet. This means that they have genes that make them react more to calorie intake or reduction than others. In the lucrative fat business, commerce is never far behind science. There is already a company – Acylin Therapeutics – that is developing what they hope is a blockbuster diet drug called a GOAT (Ghrelin O-Acetyl Transferase) inhibitor. It works via a form of epigenetics called histone acetylation, which alters DNA folding and inhibits the effects and expression of Ghrelin and in theory reduces hunger and thus obesity.[8]

We don't know if this will work yet, or whether there will be side effects of silencing the hunger gene, but it sounds worth a try. Once you have carefully and conscientiously built up your extra fat stores though overeating, over 16 genes, including novel genes like ISR1,[9] have been recently found to influence where fat is then stored,[10] e.g. under our skin or around our internal organs, on our bottoms or on our thighs. As we showed many years ago in a twin study, this varies widely between us and is genetically influenced. Hence women who can be pear-shaped like Marilyn Monroe and apple-shaped liked Dawn French or Rosanne Barr.[11] The newspapers at the time of the article carried pictures of models like Kate Moss, but they have no fat anywhere, so it's hard to tell what fruit they are – maybe asparagus.

Although these genes help us understand the mechanisms better, they don't explain the recent dramatic weight changes in the population or the differences between identical twins. Using our twins and a tiny bit of fat tissue they kindly donated to us from below their belly buttons, along with my colleagues in the MuTHER Consortium we recently found a new master regulator gene for obesity, called KLF14. This 'master' gene was already linked to diabetes and cholesterol. We found that it also influences the behaviour and expression of many other distant genes that control BMI and blood sugar.[12] The KLF14 gene is unusual in that its activity is inherited from the mother. Everyone inherits a set of all genes from both parents, but in this case the copy of KLF14 from the father is switched off and only the copy from the mother is active – an extreme epigenetic process we have discussed before, known as imprinting. So the first master regulator of obesity found just happens to be an imprinted gene. This along with the epigenetic changes to FTO[13] show that it is unlikely to be just chance that is working here, making it probable that epigenetics has a pivotal role in obesity, diabetes and other metabolic conditions.

Dorothy was the younger and chubbier of the twin babies. Although the same height, 5' 8", she is now, aged 57, considerably heavier than her twin sister Carol, with a difference of over 4 stone or 60 lb in weight.

'I've always been the bigger eater – some people have said I'm like a gannet. Carol and I watch what we eat and are careful, although I admit I have probably always eaten a little bit more than her.' They were not aware of any remarkable differences between them in diet or choice of foods. 'I think we have usually eaten the same sorts of things, which was quite healthy, as our dad had an allotment and gave us plenty of fresh fruit and vegetables,' said Dorothy. 'I have always been allergic to tomatoes and Carol to chicken.' They believe they have similar personalities, but admit that Carol is the more outgoing, and they always had a different circle of friends. 'We both do about the same amount of exercise – I swim, Carol does yoga and walks, and as children we were not very sporty generally.'

As children neither she nor her sister were allowed a bike or to explore or play on their own. 'Our father was rather overprotective because we grew up in the North of England at the time of the local Mary Bell murders,' remembered Carol. (Mary Bell was a deranged ten-year-old girl who in 1968 murdered and mutilated two young boys and had herself been badly abused.) Their two younger sisters are short and plump, while their brother is tall and slim. There have been no major health differences between the twins, except that Dorothy had an early hysterectomy aged 35, in the mid-1990s when it was 'fashionable'. 'My weight definitely started to increase a lot just after that time,' she said.

'I just think that the difference in weight between us is due to different metabolisms,' Dorothy says. Neither twin drinks much or smokes. Carol has a different view: 'I'm sure that the differences between us in minor food allergies have been important in our weight.', In her forties Carol went to see a private dietitian locally in Sussex. She had early menopausal symptoms and found her weight soaring, going from 9 to 11 stone. She was diagnosed from her story as having minor wheat and lactose (dairy) intolerance, apparently confirmed by various vitamin blood tests.

After a year of trying different diets and restrictions without success, Carol 'followed the advice to cut out all dairy and wheat products and soon improved mentally and physically and got my weight down to nine and a half stone, where it stayed.' Dorothy didn't feel the need to try the same. Carol and Dorothy's favourite food is still rice pudding, although Carol now cooks it with soy milk. Neither of them is a fussy eater, and they have few foods they dislike, apart from squid. They both seem pretty happy and content with life. They don't meet as often as they would like but talk on the phone every other day.

Identical twins like Carol and Dorothy with very different weights are actually hard to find, but as with investigating rare diseases and syndromes, the exceptions often prove more useful than the rule in helping to understand specific mechanisms. So could food allergies, as Carol believes, be the reason for their weight differences? Around one in four adults now believes they have food allergies, and nearly everyone can report being intolerant to something. This is aided by a constant bombardment of media stories fuelled by the para-scientific food-allergy business, full of bogus hair analysis, dodgy blood tests and nutrition plans. But the accurate figure for real confirmed food allergies in adults is less than 2–3 per cent. Nevertheless, some allergies, like that to peanuts, are really increasing, and external factors like peanut oil use on babies could be responsible.[14]

Food intolerance is a vague term, with many disparate symptoms caused by a wide variety of factors, and is often temporary. In Carol's case she never had any of the classic symptoms of lactose intolerance, namely nausea, bloating and sickness after eating milk, which usually leads to weight loss, not gain. But anyway who am I to disagree with a nutritionist? And the diet worked – possibly because it made her control strictly what she ate.

However, lactose (milk) intolerance and real allergy is common in kids, and is determined by a gene coding for lactase, the enzyme that breaks down the lactose in milk. This gene mutated recently in human history, about 7,500–9,000 years ago, to allow our ancestors entering Europe to drink raw cow's milk while on the trek northwards[15] without being sick, and enabled cattle to be used as sources of protein. Scandinavian and British populations have very few people without the mutation, Southern Europeans about 50 per cent, and some African tribes only about 10 per cent, explaining the differences in milk use across Europe and the world. Cheese protein is already partly broken down and so is more widely eaten, without the gene mutation. But recent studies have shown that some populations that lack the North European mutation can still drink milk, suggesting that other factors, such as other genes or epigenetics, could play a role. Clearly, if Carol and Dorothy really do differ in their response to milk then epigenetic factors must be important, perhaps switching off the lactase-producing gene. But obesity is not really a consequence of food tolerance or intolerance, but rather of food choice and intakes combined with too little energy expenditure (exercise). Could subtle differences in their food choices or intakes have made the crucial difference to their weights?

What makes people choose unhealthy foods? Well, you won't be surprised by now if I tell you that genes do play a role. Clearly having good and varied taste receptors has been important for all animals in evolution – some people naturally prefer salty, savoury foods and others sweet tastes. A single gene controlling a receptor in our tongues called PTC allows some people to be blissfully unaware of an extremely bitter compound that two-thirds of us find very unpleasant. In most foods though, which are a complex mix of sugars, fats and proteins, the genetic influence is weaker and much more complex.

We found heritabilities of around 30–40 per cent for whether people choose to eat more or less salad and vegetables, meat products, fried foods, garlic, dairy or even chocolate.[16] We know there are also genes influencing tea and coffee consumption as well as alcohol.[17] Genes even predisposed subjects to prefer or dislike a healthy food pattern of fresh and varied produce (so-called Waitrose diet) versus a traditional, i.e. modern British (fried food), diet. Some people really find it hard to eat salads and other vegetables as they find the taste bitter. Taste variation can partly explain other habits, like the non-smokers who often try cigarettes but find the tobacco too bitter to ever get hooked. This is likely to result from a combination of not only genes but also training and habituation. Nevertheless it would partly explain the mixed results of implementing public health messages of five daily portions of fruit and veg across all individuals.

Years ago, I helped set up with Sir John Sulston (the Human

Genome Project lead geneticist) one of thankfully few live and stressful TV twin experiments for the Royal Institute Christmas Science Lecture he was giving. We recruited twin schoolchildren and placed them on stage hidden from each other. We found that identical twins had more similar tastes in choosing hard or soft-centred chocolates. One pair of twins actually spat out their chocolate simultaneously live on camera. In contrast, we found that a preference for sweeter milk or more rarely bitter dark chocolate was largely due to culture and upbringing. Probably we would have seen different results in Belgium or France, where dark chocolate is the norm.

Cultural factors, habit and genes are combining in our choice of what we put in our mouths and ultimately whether we are susceptible to being vegetarians, fruitarians, vegans, or carnivores. The long-term health benefits of different diets are hard to quantify, as they are usually combined with other lifestyle differences like smoking and drinking. But even accounting for this healthier lifestyle, the religious group the Seventh Day Adventists, who don't eat meat and sometimes fish, have been found to live over seven years more on average than their Californian neighbours who were being fed the Great American Diet.[18] As little as 60 years ago in a world where food was not abundant, not many of us – whatever our genes – would have refused a nice juicy piece of meat, which was seen as a symbol of wealth and health. But the diet and food choices of our ancestors may have influenced our risks of obesity and its deadly sister diabetes.

Let us return to the Agouti mouse we discussed briefly at the beginning of the book. The Agouti gene is named after the coat colour found in artificially bred guinea pigs which are South American rodents bred originally to be eaten. They have a mix of black-, yellow- and red-banded hairs. Although Sewell Wright, a famous geneticist, had suggested as far back as 1916 that an odd form of genetics might be operating in coat colour, he was largely ignored, and it took other scientists working with mice and the Agouti gene decades later to prove it.

The Agouti gene and the protein it produces interfere with the production of the dark pigment melanin, responsible for our hair and skin colour. But it doesn't act in a predictable manner as most genes do. A litter of genetically identical pups (with a particular gene variant called Agouti variable yellow) have variable coat colours – ranging from blonde to mottled to brunettes. The blondes however in this case don't have all the fun. Compared to the slim brunettes they are obese, diabetic and prone to cancer – even though they had identical genes and DNA structure to their brunette siblings.

Researchers from Duke University performed the key experiment when they fed identical brunette mothers who were eight weeks old a special diet containing chemicals that had extra methyl donors. These included vitamins like folate, vitamin B12 and choline, commonly found in many 'healthy' foods such as leaf vegetables, liver, cauliflower, beans, nuts and garlic. These methyl-supplemented mums produced pups who, although not given any supplements themselves, were predominantly healthy slim brunettes, compared with unhealthy chubby blondes in the unsupplemented group. This was directly due to the maternal supplements switching off the genes by methylation.[19]

Many pregnant mothers around the developed world have for the last 20 years taken folic acid vitamin supplements to reduce the chances of spina bifida and birth defects.[20] In the US many foods, including bread, are routinely supplemented with it. While this seems to work in reducing birth defects, few realise that it could have other unknown effects on their genes that could influence them and even their grandchildren. Most women are probably also deficient in the B complex vitamin choline, although we know little about this essential nutrient and others like its derivative betaine that could alter methylation or gene function. Other rodent studies like the Duke experiment have previously shown that maternal diets and nutrients are also important for the offspring's development. When rat mothers are either calorie- or protein-restricted while pregnant[21] they will usually produce smaller-sized granddaughters that have a form of diabetes.[22] This occurs via epigenetic mechanisms.

We know that mothers can have direct epigenetic effects on their offspring's genes via normal (somatic) body cells. We saw this when discussing the altered behaviour in the licking studies of young pups. Clearly a pregnant mother has a close direct contact with her fetus for many months before this. Strangely she is also in contact with her future granddaughter, since, as in Russian dolls (or the film *Alien*), the fetus is itself developing primitive egg cells containing the seeds of the next generation. So any effect of environment on pregnant females can affect the next two generations of females directly via the DNA in the somatic cells. This means that studying only females makes it difficult to be certain that it is through the genes and the germline cells that the effect is being passed on. That is why we need the dads.

Males – as so often – are simpler. They have only a brief encounter with the future mother for sex and then usually little contact with their offspring. These are rodents we are talking about. The only link of these fathers with their future children and grandchildren is via their sperm. A recent study has shown for the first time how a father's diet can affect his daughters' health. When they fed male rats a high-fat diet, the outcome was not surprising – just as in humans fed a McDonald's diet, the dads' body weight and body fat increased, and for good measure they developed early diabetes and resistance to the hormone insulin.[23] Unexpectedly, however, although their daughters were not fat like their dads, as they got older they developed a diabetes-like condition of abnormal glucose tolerance and insulin secretion.

They also found that the genes being expressed in the cells that make insulin (in the islets of the pancreas) in the daughters were abnormal, altering several gene networks and cellular pathways. This told them that the fathers' high-fat diets altered the development of their sperm, which then led to epigenetic reprogramming of these cells, promoting an adult-onset diabetes in the daughters. A subsequent study looked at mice whose fathers were fed a low-protein diet (10 compared with the normal 20 per cent) and increased sucrose. The 'vegetarian' dads had modest but widespread methylation changes in the liver of key genes that influence not only weight, but diabetes and lipid production (cholesterol).[24] These gene changes are quite different to the rare genetic mutations that can occur in sperm due to anti-cancer drugs and some toxins.

Willpower and marshmallows

Jenny and Florence looked very similar but were always fighting. Until we first tested them at the age of 40 they had no idea they were identical twins, as their mum had always told them the midwife was sure they were non-identical. They were born with similar weights of 6 lb, 50-odd years ago in India, and spent most of their childhood in English boarding schools with their notoriously bad food. At school they both played sports but both slowly gained weight, eating snacks of bread and jam. Just before they left school both twins were of average height at 5′ 5″. Jenny was around 11 stone and her sister Florence 12 stone.

Florence, who was always naughtier and had missed a year of school while pretending to be homesick, said: 'I decided one day after talking to my brother, who was thinking of dieting, that I was tired of being overweight and decided to do something about it myself.' She went on a strict diet for two months of apples and cheese, she became anaemic and her periods stopped as she lost 4 stone in a year. She never gained weight again. She impulsively eloped and moved to Illinois in the US and married a military man and became a firm vegetarian, cutting out most carbohydrates. She is still susceptible to the odd small piece of dark chocolate 'because it is full of antioxidants'. Florence now weighs just over 9 stone and Jenny, who married an Englishman a few years later, now weighs nearly 15 stone. Jenny introduces her sister as 'my thin twin'.

'The reason I'm the fatter,' Jenny (who is a pharmacist) says, bluntly, 'is because I eat more. I snack and comfort-eat, my sister doesn't – she is very controlled. I think my life is less stressful, happier and more relaxed than Florence's, so I don't feel the need to control that part of my life. I have tried a few diets half-heartedly but they have not worked. I know I could lose weight by eating less and doing more exercise, but for me, it's not a priority.' 'Although I'm the fatter one,' she says, 'I think I'm happier.'

Jenny has twin 15-year-old sons, who are both of normal weight and tease her. However, both the children of the thin twin, Florence, are, unlike her, seriously overweight. Her daughter now weighs over 300 lb (over 24 stone) and has needed gastric surgery for obesity. Her husband is travelling most of the year and she works locally as a hairdresser.

Florence doesn't agree with Jenny's theory: 'I've had a tougher life, but I feel happy in myself. I'm healthy and Jenny has health-related weight problems she doesn't admit to, like joint and back problems. Once I've decided to do something, I stick with it, whatever – like my thirty-year marriage.' Both twins do agree they are strong-willed and slightly obsessional. They both easily gave up smoking. They are both a bit crazy about tidiness around the house. Their father was also strong-willed. He too was obese as a child and adolescent, having been force-fed by his Greek grandmother. He managed to slim down though, unlike his four fat brothers.

Clearly genes, parents and diets are important – but so is willpower. Although both twins had the same opportunities to diet, only one took them, possibly leading to different effects on their future children. With knowledge and hindsight they can now blame in part the food culture of their Greek grandmother. Willpower is hard to define, and it is even harder to identify which part of ourselves, our subconscious or our brain, drives it. We do know however that possession of it is a good guide to success in life, as shown by the now classic marshmallow experiment.[25]

The experiment was designed by Walter Mischel of Stanford University in 1976. A group of four- to six-year-olds were placed in a plain room at a table with a single marshmallow on it. They were told if they waited alone for 15 minutes without eating it they could have a second one when the adult returned. Most kids waited for a few minutes, then ate it. But about a third waited patiently (many squirming in their seats) for their reward. This form of exercising willpower is known as delayed gratification. When he followed them up 14 years later he found that according to their parents and their school exam results, they had a marked superiority in most domains over the instant-gratification kids. The same group also found that willpower can be altered and trained, so that kids who imagined their marshmallow as an inanimate picture could more easily wait the 15 minutes.[26]

I haven't found any direct tests of the genetics of willpower in twins, but one test we did live for the BBC documentary pitted twins against each other for how long they could hold their arms underwater in a bucket of ice. Some were wimps who lasted a few seconds and others macho fools who competed against each other for ten minutes, nearly fainting in the process. The identical twins performed pretty similarly to each other, probably due to genetic and cultural similarities in both willpower and pain sensitivity.[27] Other experiments and extreme examples such as David Blaine have shown that willpower, in terms of both its strength and duration, is clearly plastic and amenable to change.[28] It doesn't seem to matter what type of trivial self-control you practise (such as sitting up straight): even after a few weeks it helps willpower in much wider ways. Websites have even been set up to help you.[29] So willpower and delayed gratification are probably both genetic and trainable, being modifiable by your own efforts and perhaps those of previous generations.

The dieting cycle

Moving from willpower, let us just think a minute about diets and the foods we choose to eat. In the 6 million years since we parted company from other vegetarian primates we have slowly evolved to be omnivores, giving us greater versatility and less time spent foraging. We believe early humans ate a very wide variety of foodstuffs – essentially anything they could find and identify as edible.

In the thirteenth and fourteenth centuries, when men were still apparently fit and relatively tall, we know that the average Englishman's daily diet was dominated by grain. This was 75 per cent from seven types of grain (wheat, rye, corn, barley, maize, oats, millet). He would eat this as a large daily loaf of bread, along with thick soup (pottage), and to help it down and stave off thirst a gallon (about 8 pints) of weak ale – or more if he was lucky enough to be a member of the clergy. This was supplemented with peas, beans and other fruit and vegetables from the small gardens that most people kept. Fish was eaten on Fridays and during Lent. Meat was a rare treat, though more common for the rich. Even back then, they had medieval dietary gurus telling them never to mix different foods such as salads and cold meats or eat them in the wrong order: 'never ever eat pears last lest it lead to a bad mix of the body humors and rotting and putrefactions in the intestines'.[30]

Over the last 20 years, coinciding with the obesity epidemic, dieting has become a global obsession as thousands of diet gurus sell millions of their unique recipes to the overweight, all promising the ultimate cure. These offers to help usually come with their own clubs, websites and merchandising. There are over 60,000 titles listed on Amazon with diet in the title – over five times the figure for happiness – with an estimated 25,000 different diets.

From the calorie-counting diets (such as WeightWatchers, Slim-Fast, GI and F-plan), to the high-protein (Atkins and South Beach), to the food-mixing (Montignac), vegetarian and vegan (raw food, China Study), Historical (Palaeolithic), utensil-orientated (Le Forking), model drill sergeant (Skinny Bitch), calorie restriction (CR way), to the recent craze, a complex mixed model low fat (Dukan Diet), consumers confront an impressive choice of solutions. There are even Breatharian diets and websites for those gullible few who believe they can live on air. While many certainly work short-term, like the Atkins high-protein no-carb diets, by making you feel slightly sick and less hungry, they are unsustainable long-term. They often make you ill with side effects like constipation, halitosis, gout and kidney problems while they manage to obsess you even more with the thing you are most trying to forget – food.

Generally books are written by unqualified self-declared experts we put our faith in, like the famous Dr Atkins, who against his predictions died young and overweight of heart disease. In most countries anyone can call themselves a nutritionist and buy themselves a mail-order degree. Even within each of the specific diet camps, the so-called experts can completely disagree with each other, as with the feuding in the raw food world over whether you should eat only fruit, only vegetables, or liquids, have mega-supplements or not; or those that fight over the best types of fats; or whether fish protein is good or bad; whether you should eat meals in small quantities by grazing, or eat a big lunch, or a big breakfast, eliminate the evening meal, or avoid 'bad' white carbs. The reason for the disputes is that there is no serious proven science in terms of any long-term human studies behind any of them. None of the profit-making dieting gurus or their companies support any proper research of their diets in case it provides the wrong result.

Surveys suggest that most people have tried a diet, with only a few per cent succeeding long-term. As we are all getting fatter – current estimates state that in the US men and women are gaining nearly a pound a year, mainly due to intake of potatoes and crisps[31] – we can conclude that most diets clearly fail. It's not all bad news though. The diet industry is thriving, and like us grows ever fatter. In the US alone it is worth an estimated $50–300 billion annually.

Some of this failure to control weight may be due to diet cycling. Not some new type of diet that you eat while doing a spin class (although that has potential), but rapid losses in weight followed by gains (also-called yo-yo dieting) that can reset the appetite genes epigenetically in the somatic cells of the brain.

After having less food than the brain is used to, our sensitive defences imagine we are in a famine situation and then reduce energy use in everything non-essential. Unfortunately this slowdown includes our metabolism and digestive systems, which translates into burning fewer calories. Each time we diet successfully, studies show that our body tries 'helpfully' to return us to our previous maximum weight.[32] Our hormones still provide hunger signals even a year after successful weight loss.[33] This explains why follow-up studies of champion weight-watchers who lost weight dramatically found them all overweight again five years later – well after the publicity shots were taken.

Twin studies conducted over 20 years ago, and which might be considered unethical now, overfed normal young male identical twins an extra 1,000 Kcal per day for 80 days. There was a wide variety of weight gains, with an average of 8 kg. Some individuals seem to be able to get rid of the excess calories more easily than others. One study has suggested this is by subconscious fidgeting and an inability to sit still (technically called NEAT – non-exercise activity thermogenesis), which may be a genetic trait.[34]

What about response to diets?[35] Regardless of whether they work long-term, weight-loss responses among overweight individuals to a fixed-calorie diet also vary greatly.[36] Another twin study looked at a strictly controlled weight-reduction programme of around 400 KCal over 28 days supervised by tough staff in a Czech hospital so that they couldn't cheat or give up. The volunteers were obese female identical twins aged 30–45. No surprises – everyone lost weight. They started with an average weight of 94 kg and lost an impressive 9 kg. However, some lost only 6 kg and others over 12 kg – while on exactly the same 28-day diet. The identical pairs matched each other very closely for weight loss, with a strong though not perfect correlation of 85 per cent.

While most explanations of differences to date concentrate on genetic effects in body-fat composition, it is quite likely that epigenetics and past experience is also a major player. Overweight Canadian volunteers kindly donated fat biopsies during a six-month calorie restriction diet. The fat cells of the slow weight losers showed significantly altered DNA methylation and expression of 644 genes compared with the rapid losers.[37] Many of these genes were known candidates and highly expressed in the brain – again suggesting that this is the key body organ involved. Quite when and how these methylation changes happen to the genes is unclear.

One study from Southampton took DNA from the umbilical cord at birth and compared it with how fat children are at age nine. They found that certain candidate genes (such as the retinoid receptor) are already methylated in those who become chubby later. Interestingly, these kids had mothers who had very low carbohydrate intakes in early pregnancy and relatively higher fat and protein consumption.[38] Although we can't yet link this directly to adults, it shows the importance of reversible methylation changes related to the maternal and likely the grandmother's diet. This suggests that we may in the future be able to reverse some of the bad influences or genes with targeted nutrition.

Why do Californians want to live like worms?

Some people are not just interested in diets to lose weight – they want to live for ever. For the past 18 years Paul McGlothin and Meredith Averill have shared an unusual life in upstate New York. They wake at dawn, eat a hearty breakfast of vegetables and fruit and do some work. At 2 p.m. they have a lighter meal, usually some very dilute soup with a tomato floating in it. They consume no sugar, no alcohol and only a few carbohydrates. For supper, while their neighbours are settling down for a pot-roast dinner, Paul and Meredith skip food and go for an invigorating one-hour run in the woods. He consumes 1,900 calories a day, she 1,600 (consisting of 9 per cent protein, 59 per cent carbs, and 32 per cent fat). This is about 30 per cent fewer calories than recommended intakes of semi-sedentary people. They eat between one and three meals a day, warding off hunger with tiny courses that they call 'teaser meals' and by practising 'savoury meditation'. They don't eat after 3 pm and aim to be fasting for about 16 hours.

They believe that by following their diet they will be around to see scientific developments that will prolong their lives even further. 'Stem-cell research, will soon make it possible to replace worn-out body parts, and we await possible immortality, We know that time is coming and it'd be great to be a part of it,' McGlothin enthuses. They are the ultimate delayed-gratification experts, resisting all temptations, and have been likened to religious zealots awaiting the Rapture.

Averill, at 5' 4" and 105 lb, is pretty lean, but McGlothin, 5' 11" and only 103 lb, looks extremely skinny, although they say they try to keep their BMI to 18, which is just above clinical anorexia levels. They believe this calorie restriction (CR) is the key to a longer and healthier life. Through their best-selling book *The CR Way* they are converting many other like-minded folk and say they have 100,000 followers worldwide. Their website contains a range of memberships from up to $395 each, plus CR sunglasses, glucose strips and other accessories and information.[39] They are apparently the leading calorie-restricting researchers in the world, and although their website links them to Nobel Prize winners like Elizabeth Blackburn, like most other nutritional gurus they never publish any research in recognised journals.

The CR followers base their philosophy and lifestyle on scientific research on animals. The first studies, carried out over 70 years ago, showed that mice on a reduced diet tended to live longer. More recent studies of yeast and worms have shown the same phenomenon. Worms are one of the favourite tools of longevity researchers, since you can totally control their diet, they don't live too long and now you can alter their genes. If you put them in a near-starvation state they respond by shutting down the non-essential systems: they stop having sex, avoid movement and live seven times longer.

The CR folks hero-worship the work of Leonard Guarente from the University of Washington, who showed the key importance in this process of the sirtuin gene family in animal models of ageing.[40] The sirtuins are molecules that can act epigenetically by unfolding and exposing the DNA loops to allow more gene expression, and have a wide range of roles in disease.[41] Sirtuins are switched on and off by certain foods and nutrients. In experiments when animals are given sirtuin activators to mimic caloric restriction, the lifespan of worms, flies and fish – but not rodents – can be extended.[42]

There is now some doubt over whether this gene family works the same way in humans. We have recently looked at the common gene variations in over 20,000 adults, and found some important genes associated with biological ageing in humans (as measured by telomeres), but none of the seven genes of the sirtuin family were even weakly implicated. Other recent studies now suggest that the original sirtuin findings in animals may have been errors.[43] Other recent studies suggest that changes to sirtuin genes (Sirt 6) can extend longevity in rodents, but only in males.[44] More worryingly, others are showing that different genetic strains of mice (called G93A) when calorie-restricted actually age faster, paradoxically due to increased oxidative stress.[45]

While the CR community are probably correct in thinking that a diet rich in fruit and vegetables and lots of exercise will make you healthier,[46] starving yourself for 16 hours a day may have the opposite effect. As seen during the Dutch famine of 1944 and the later Chinese famines, extreme food deprivation can have unpredictable epigenetic effects on the next generation – though if you have no libido or regular periods this may not be a problem. We have shown in large studies that lifestyle factors like stopping smoking, avoiding obesity, and moderate exercising can retard biological age by five years, and in contrast chronic stress is harmful.[47] Time will tell whether the CR way of cyclical starving and exercising is perfect for the body, or for some people too stressful. Even if it worked, not all of us would opt to live for 150 years if we had to look forward to the lifestyle of a celibate worm.

If starving and running for years isn't your cup of tea, how about a healthy glass of wine? There is increasingly good evidence from various animals including rodents that resveratrol, a plant extract that is found in wine, is a sirtuin activator and can mimic many of the biochemical effects of calorie restriction.[48] It increases mitochondrial activity (mitochondria are the body's energy cells), appears to particularly help obese animals improve their insulin sensitivity, and in lower animals such as yeast and worms can improve lifespan. It appears to do this by switching on sirtuin genes which in turn de-acetylate and activate other gene pathways and can be mimicked by some drugs (PDE4Is).[49]

Studies have generally shown wine to be beneficial in humans in reducing heart disease and some cancers, and may explain why the French have relatively little heart disease. However, only small amounts of resveratrol are contained in wine, so that unless you are dedicated and drink several bottles daily, other nutrients including alcohol itself or procyanidins could be responsible.[50] You can also obtain resveratrol from peanuts and chocolate, which are not hugely popular in the CR world. White wine and grape juice strangely contain very little, and red wines even from the same region vary widely in their content. Some studies suggest that the Pinot grape or Super Tuscan wine varieties may be slightly superior. Sadly, it looks like a case of having to try them all.[51]

Vitamins, grandparents and Renaissance

We have seen how obesity and diabetes are influenced in part by our genes and by what we choose to eat and – critically important – by what our parents and grandparents ate at critical times of their lives. Epigenetic marks placed in one generation will influence gene expression in the next few generations. This is a short-term evolutionary adaptation to major environmental challenges, such that adverse exposures in one generation (like famine, cold, or toxins) act as warning signals for the next generation. Subsequent generations will then be primed to do something different from the expected. They might react by getting fatter or thinner or becoming more active or sluggish. As the warning signal wanes, provided no other shocks occur, the next few generations will slowly return to responding more 'normally', meaning in less extreme ways.

Nature appears to have given us a parallel evolutionary survival plan allowing us to adapt quickly to adversity. But evolution over previous millennia could not prepare humans for our current unprecedented glut of food and for sedentary lifestyles. While some researchers have identified individuals at high genetic risk of diabetes and successfully managed to get them to reduce their risk by weight reduction and exercising more, this is not easy in the long term.[52] The challenge is to understand the biology of epigenetics properly so that we can intervene in a targeted way and minimise or reverse the effects on ourselves and our children.

One way to reprogramme our genes epigenetically might be through food and mineral supplements, as we have seen with folic acid in rodents. The unregulated mineral and vitamin supplement business is worth over $50 billion and claims it is the route to unlimited good health without the need to eat boring fish, fruit and vegetables. However, despite naïve but clamorous media coverage, and wild claims by the industry, very few supplements, although often promising in theory, have any proven value. Good examples, as pointed out repeatedly by the medical investigative journalist Ben Goldacre, are the use and abuse of the antioxidant supplements (vitamins A, C, E, beta-carotene and selenium) to halt heart disease, cancer and the ageing process.[53] The companies and the media prefer to overlook gold-standard impartial reviews that offer meta-analysis of all the properly conducted randomised trials – trials that show these remedies just don't work.[54]

One such review looked at 67 trials involving 230,000 people, which showed that antioxidant vitamin pills do not reduce deaths, and in fact may actually – as in the case of beta-carotene and vitamin A and E – increase your chance of dying. Trials of these supplements had to be stopped early because the group taking the 'healthy supplements' developed more lung cancers than the untreated group.[55] The results of the review were not popular with the industry or with scientific experts like Sir Cliff

Richard, who complained of bias and a witch-hunt.

Other examples of hype are the persistent claims by manufacturers that fish oils can improve children's IQ and school performance, despite exposure of many of these claims as fraudulent, and the makers' denial of negative independently performed trials.[56] Other mainstream vitamin supplements like calcium, given for bones, are also being seriously questioned and may have unhealthy long-term side effects like heart disease when given as a single artificial dose and not as part of natural foods. Other vitamins will likely come under similar scrutiny.[57]

Whether our grandchildren as adults will be on average fatter or thinner than we are today is hard to predict. The Swedish human famine data we discuss more in the last chapter suggest that after the grandparents overate as adolescent children (the way most of us do now) their grandkids were more likely to have diabetes and so obesity. As famines are unlikely, my guess is that even if calorie intakes don't rise further, our grandchildren will on average get slightly fatter – unless we do something about it.

If my description of diets, vitamins and supplements for keeping you thin, healthy and long-lived has left you confused, you are not alone. I am certainly not going to offer any specific advice (unless you sign up long-term to my very expensive website and merchandising), other than to give these words of wisdom: eat as much fruit and vegetables as you can and do plenty of exercise. As for choosing between caloric restriction and a glass of wine – personally, I'd go for the wine. You never know, your grandchildren may one day thank you.

9

'THE CANCER GENE'

Autism, toxins and babies

At the age of just 23, Kristin mentioned worries about a lump in her breast to her mum and twin sister Maren on a trip to Barcelona. Upon their return home, Kris went to see her GP, who examined her and said that it might just be a hormonal reaction to the contraceptive pill. They briefly discussed breast cancer – Kris's grandmother had had the disease in her thirties but lived till 75 – but the GP was not concerned and prescribed evening primrose oil. The pain continued, and six months later after a long work trip to China, Kris returned. A different GP, without examining her, suggested that she change her contraceptive pill – which she stopped, to no effect. At last, several weeks later, at the insistence of her mother, Kris returned to her original GP and insisted on a referral to the local hospital, where the diagnosis was eventually made. It was breast cancer.

'It was a real shock for me but also very hard for me to tell Maren, but I knew it would be even harder for her to hear. But she needed to be tested too,' she said. Maren quickly had an MRI scan, which came up clear. Further tests brought more bad news: Kris's tumour was already very large and had spread to her spine. As she put it bluntly: 'I've got stage-four cancer. There is no stage five.'

Three years on, Kris amazingly is still very much alive and has now had chemotherapy and radiotherapy to shrink the tumour, followed by a mastectomy. Having been totally bald when I last saw her, she has now regrown her hair and is having cement put into a vertebra of her spine damaged by radiotherapy and the tumour. She takes Tamoxifen, the anti-oestrogen cancer drug. 'One of the worst things about being bald and sick was not being recognised as twins any more – that was a real low point – but I'm much better now, and look like Maren again.'

Kris and Maren are the daughters of Anglo-German parents. They spent the first half of their childhood in Germany, but moved to the Midlands after their parents divorced in the mid-1990s. At school the tall skinny blonde Germans (as they were known) were very close, sat in the same classes, studied the same A levels, and achieved the same top grades (three As). They had very similar upbringings and environment until the age of 19, and even travelled round Australia in their gap year together. They then chose different universities. They kept in close contact through student life and spoke every day.

'I did think: "Why her, why not me?"' says Maren. 'But then I realised that had I been asked which one of us was more likely to get cancer, it was always going to be Kris. She was always the one with most health problems. She alone had headaches and was homesick, she had her appendix and wisdom teeth removed, while I still have mine.' 'I guess I was the slightly weaker more anxious one, who got more stressed,' agrees Kris. 'My sister was always a little bit stronger.'

Apart from this, there were no major clues to why Kris was affected and not Maren. They weighed the same at birth – 7 lb with no complications. They both grew well, had their first periods late – at 14 and 15 – had their first boyfriends at 17, and both had irregular periods. Maren didn't get on with the contraceptive pill and stopped after a few weeks, whereas Kris was on it for two years. They both had the same identical 'Wurst-und-Kartoffeln' (sausage and chips) diet for their early German years, then continued with their mum's fish fingers. They are now both non-dairy vegetarians since the diagnosis.

Genetic tests performed in Oxford have established that Kris is not a carrier of the rare breast cancer gene mutations called BRCA1 and BRCA2, which is a relief to Maren, who now has yearly scans. Kris is positive and upbeat: 'I could let those questions about my cancer get to me and stop me from doing things. But I can't see into the future and I definitely can't live every day clouded by the thought of death. So the only answer is to live – and live well.' Kris has set up a charity called CoppaFeel, which is educating young women about self-screening early in life, to reduce delayed diagnoses like hers.

Most people, if asked, would assume that breast cancer is strongly genetic. However, although it runs strongly in some unusual families with rare gene mutations such as BRCA1 and 2, it is overall only around 25 to 30 per cent heritable. This explains why identical twin sisters, who share the structure of their DNA, are often differently affected by breast cancer, as with Kris and Maren. The disease also gets less heritable with increasing age, so there is hardly any gene influence when women are over 70 years old, when the majority of cases occur. Although many think of breast cancer when we think of heritable diseases, breast cancer is in fact only about a third as heritable as a common but dull disease like back pain.[1]

When we say that breast cancer is overall only about 25 per cent heritable, obviously it doesn't mean that genes aren't involved – in fact they are crucial in the cancer process and geneticists have discovered at least 10 different sub-types depending on the genes involved[2] – it's just that these changes or genetic risks are only weakly passed from one generation to another. Maren now has less than a 50 per cent chance of developing Kris's form of breast cancer, whereas if Kris had one of the rarer BRCA1 or 2 genetic forms of the disease she would have a nearly 90 per cent chance, as these forms are much more heritable.[3]

Kris's GPs would never have seen a case in someone of her age, whereas they commonly see breast-tissue changes due to the contraceptive pill, so Kris was unlucky in that regard. We now know too that women with firmer, high-density breast tissue (which is mainly genetic) have higher cancer risks and are also harder to examine and feel lumps in. Breast cancer is very rare in the young and produces only around 20 cases a year in the UK in women under age 25. It then increases slowly with age. Eighty per cent of breast cancer is diagnosed over the age of 50, when most routine screening starts. It affects around one in eight women in the UK and USA, and rates have generally been steadily rising over the last 30 years, with numbers doubling in the UK.[4] Although there is a real increase, some inflation comes from better diagnosis and screening and a recent trend of labelling early reversible changes (called cancer in situ) as definite cancer. Breast-cancer rates also vary widely across countries. Northern Europe has four times the rates of Africa and Asia, although there have been recent dramatic rises in Japan, and in cities in China experiencing social and economic changes, which can't be explained by differences in screening.[5]

What exactly caused Kris's cancer – the most common cancer in women – that the odds say will not affect her identical sister? Epidemiologists have thought they know most of the associated risk signals for several decades. They include a number of hormonal factors, including an early menarche (start of periods and puberty), being older at first pregnancy, and a late menopause. All of these factors, as well as not breastfeeding, increase the number of periods and reproductive cycles a woman undergoes in a lifetime. The contraceptive pill (invented mainly by Carl Djerassi 60 years ago and ironically put into clinical trials by a staunch Catholic, John Rock) strangely has both good and bad effects for cancer.[6] On the plus side, they reduce the total lifetime number of normal periods a woman has and mimic pregnancy, which is protective. This 'natural' effect was why John Rock thought it would please the pope.

However, the hormones, oestrogen and progesterone overstimulate growth of breast tissue, which is a risk factor. The two effects more or less cancel each other out, leading to only a small – 15 per cent – increased risk which luckily disappears quickly on stopping taking the pill. Natural oestrogen is important throughout life. Fat tissue stores oestrogen and obese women have a 30 per cent greater risk of breast cancer.[7] Women who as newborns were chubbier than average also have an increased risk. This is probably due to higher levels of oestrogen in their mother's blood. Those whose mothers suffered high blood pressure in pregnancy had reduced risks and low oestrogens.[8] So changes to older childbirth, higher birth weights, reduced breastfeeding and rising obesity all influence early oestrogen levels and could explain some of the recent increases in the cancer.

As usual there is controversy as to the effects of dietary factors. The strongest evidence seems to be for fat intake: a meta-analysis of 45 studies reported that higher total fat intake increased breast-cancer risk by 13 per cent, while other large studies implicate saturated fats.[9] Phyto-oestrogens – plant compounds that mimic oestrogen – have been extensively studied and marketed as a 'natural' menopause cure. The main type consumed in Western diets are lignans, found in seeds, broccoli and strawberries, which if you eat a lot could theoretically reduce your risk by 15 per cent.[10] The other type of phyto-oestrogens are isoflavones, found in soybeans, common in Asian diets. As

HRT has gone out of favour because of its breast-cancer risk, many women are taking unlicensed 'alternative' medications containing soya isoflavones, in particular genistein. Urged on by the media and marketing, these are perceived as 'natural, safe and protective' against cancer. The data show otherwise. When given to young animals genistein seems to protect against later development of breast cancer, but in older animals it will actually accelerate tumour growth.[11] Clearly, as the actual doses taken are unknown in natural products, women who choose the tablet and not the plant are taking a risk.

Epigenetic gang wars

The last few years have seen a revolution in our view of cancer. Till lately it was thought that single mutations in a 'cancer' gene usually started the process. This is unlikely. Epigenetics is now understood as the key to the cancer-forming process, by ensuring the survival of the abnormal cancer cells.

When researchers look at the whole DNA of cancer cells, they find that generally they are under-methylated – enabling many genes that are normally suppressed literally to run wild. This leads to a general instability of the chromosome that can lead to gene mutations and also to a breakdown in the imprinting system by which (in key growth genes) one copy is normally turned off. In parallel a few DNA areas show the opposite: they are hyper-methylated and the genes suppressed. These genes are the body's built-in protection system, the tumour-suppressor genes that keep the DNA under control. One of these is BRCA1, which when the gene is faulty causes a highly genetic rare form of breast cancer in the majority of women who carry it.

A group from Johns Hopkins University in Baltimore have

shown that, contrary to the idea that cancer was often random, there are amazing similarities in methylation regions among different cancer cell types, which suggest a common epigenetic mechanism ensuring these cells' survival at our expense. Thus each group of cells is perfectly placed to divide and grow rapidly and have a flexible system of cellular evolution. This allows them to adapt to new surroundings when they spread elsewhere – 'metastasise' – and to fight off a lack of blood supply and chemotherapy drugs.[12] To understand the epigenetics of cancer, let's look at another pair of twins, and how they and their cancer cells fought each other for survival.

Fifty-five-year-old Heather and Judith were unlucky enough to be identical twins who both had breast cancer. Heather was diagnosed 20 years ago, and Judith ten years previously. 'We've done everything together,' said Judith, a retired teacher. 'We had the same childhood illnesses, so in a strange way it does not surprise me that we both have the same illness now.' After successful treatment they thought all was well, until Judith went on holiday to Bermuda last year and noticed some worrying signs. On returning she found out she had a recurrence. She had to tell her sister – who within a few weeks was told exactly the same thing.

Both twins went through the chemotherapy process and further tests and scans together in Manchester. Unfortunately Heather's scans revealed extensive liver metastases. 'The doctors said this was likely to be terminal.' As Judith started to respond well to her treatment, the new but expensive drug Herceptin, her sister sadly did not, and deteriorated. 'It's the worst blow you could be dealt, but seeing Heather so happy for me makes me appreciate life and want to live every moment for her.'

Heather, a retired nurse, said: 'It was wonderful news when

she told me she was okay and in remission, even though I knew it would be tough for her to tell me. It's strange, but it felt like a relief as much as anything, as I knew she would be there for my son when I'm gone.'

There is a history of breast cancer in their family. Their mother Sheila, now 81, was diagnosed with the disease 13 years ago and their aunt at 37, though both survived. But like the younger twins they did not have a rare cancer-gene mutation. 'We are unlucky but feel philosophical – I've had a good twenty years of life since my first diagnosis and have seen my son grow up. I am grateful for every day I get with my wonderful family.'

Two years after her re-diagnosis and 22 years after the original tumour, Heather died in the hospice – the one the twins had been raising money for.

Breast-cancer response to treatment, as in these twins, is very unpredictable. Scientists are only just beginning to understand why. The drug Herceptin, which both the twins took, is effective only in blocking expression of the cancer-promoting (HER-2) gene which 20–25 per cent of tumours produce. In these cases it usually increases survival and remission. However, it has a price: it costs £20,000 for a single year's supply in the UK and over double that in the US. As we saw with Heather, it doesn't always work: the tumour mutates to avoid being killed as other abnormal genes are switched on or off. While it can be used with other anti-cancer drugs, it looks again as if epigenetics is crucial in giving cancer cells protection against chemotherapy and allowing them to metastasise. The Sloane Kettering Cancer Center in New York can now predict the early response to these drugs by looking at the methylation of tumour-suppressor genes in the biopsy.[13] These profiles are likely to be used routinely in the future in clinical practice.

The fact that cancer cells use epigenetics to quickly adapt

and mutate is also their Achilles heel. Because epigenetic changes are temporary they can be reversed. Epigenetic anti-cancer drugs have been rapidly produced, and there are at least five now available on the market in the US including Vidaza, Dacogen, Zolinza and Istodax.[14] At least another 30 are in development. Although used so far mainly in leukaemia, they seem to be working well and are better targeted with fewer side effects than conventional anti-cancer drugs.

These drugs target and try to unblock (unmethylate) the body's natural defence genes, allowing them to do the natural suppressing job that the cancer cells are preventing.[15] Other clever mechanisms have been found called TET proteins, which add to methylated genes a marker (a hydroxyl group) that then signals for the whole gene to be repaired and reset in fresh unmethylated mode.

These are exciting new prospects,[16] but the field is moving fast. Even more specific targets are also being developed. In breast cancer the tumour, in order to metastasise to bone, has first to epigenetically suppress the E-cadherin gene (responsible for cells sticking together), then once the tumour is in its new site it has to reactivate the gene so it can feed off the local blood supply. Drugs are being developed that block this, and so would stop tumours spreading.[17] One German company has developed an FDA-approved screening test in blood for colon cancer targeting a single gene, SEPT6, that changes methylation.[18] We have some encouraging early results with colleagues in Barcelona exploring 20 pairs of identical twins discordant (i.e. unlike each other) for breast cancer where we find some cancer genes are methylated in one twin and not the other.[19] Several interesting genes are regularly emerging as clearly different in their methylation patterns. These are likely to be valuable biomarkers and possible drug targets.

If we go back to the accepted hormonal risk factors for cancer for a moment, could some of these risks be passed on across generations? Two studies suggest they could. The first comes from recent studies in pregnant rats fed oestrogen, which had daughters, granddaughters and great-granddaughters with more breast tumours.[20] The other evidence comes from humans who took diethylstilboestrol (DES), a synthetic oestrogen medication that is famous for the wrong reasons. Human exposure to DES came partly from supplemented cattle feed but mainly from its medical use. From 1947 to 1971, DES was always a controversial drug. Without good evidence, it was given to pregnant women under the mistaken belief it would reduce the risk of pregnancy complications and miscarriages. In 1971, DES was shown to cause a rare vaginal tumour in girls and women who had been exposed to the drug in the womb. After much fighting and delay, it was eventually withdrawn.

For the women exposed to DES, commonly referred to as 'DES daughters', the problems didn't end there. As well as getting these rare tumours in childhood they also had an increased risk of breast cancer 40 years later. The risk even increased for women over 50. Current research is now looking at the third generation, the grandchildren of women who were given DES during pregnancy. Results so far are worrying and show they have less regular menstrual periods and reduced fertility, with a possible increased risk of ovarian cancer. All this is likely to be due to epigenetic change across generations.[21] Giving rodents DES prenatally produced the same breast cancers in the daughters as in humans, but here, interestingly, the timing is important. When given after birth DES has no effect, or may even be protective. Although not yet proven, this is highly likely to be via epigenetic modification of the oestrogen receptor genes at a critical stage. Like most epigenetic mechanisms there is considerable variation in the effects, so not all the DES daughters or granddaughters have problems.

There is a group of chemicals (called hormone disruptors) that resemble oestrogen and can mimic and influence the delicate hormone systems. They can be natural like soya beans or resveratrol from wine, or synthetic like DES or fungicides. These chemicals can affect any pathway controlled by hormones and produce defects of early development.[22] We are slowly realising how many different ones there are (we know of at least a hundred) and how frequently they appear in the modern environment. The fact that they are so widespread and can have effects at very low dosage that can occur 50 years later or even in a different generation could be a major reason why they haven't been spotted before. These hormone disruptors could be a crucial part of the puzzle not only of the increase recently in breast cancer, but also of other mysterious recent increases such as asthma, allergies, heart disease and even autism, which we will explore further.

Autism, savants and game shows

Flo and Kay are identical twins with a special talent. They can recall the day of the week for any date in the last 100 years, or when given the song title can provide the artist, date and record label of every music recording of the 1960s, 70s and 80s. They can also remember the time and date of every major US event that happened in their lives from about the age of ten. This means for any particular date you name, they can recall the weather, recall what they each had to eat, and the tie and shirt colours of their favourite TV presenters. They are unique identical twin autistic savant sisters.

On the autism spectrum, Flo and Kay fall right around where

Dustin Hoffman's character in *Rain Man* does – and they sound just like his character. They have problems socialising, can't cook for themselves, can't drive and aren't able to live independently – yet are in some aspects technically geniuses. Unlike many autistics, most of whom have no savant abilities, they have a quirky sense of humour and are passionate about music, which they like to see live. However, and classically, they avoid eye contact, handle emotions and communications poorly and have a number of obsessive behaviours. The twins are exceedingly rare, as they are female (four to five times more males than females are diagnosed with autism), they are prodigious savants (the highest level), and are strikingly similar – their abilities and disabilities appear as identical as the way they look. They have been inseparable since they were born, and still share a bedroom.

In 1996, when the popular long-running daytime US quiz show *$100,000 Pyramid* was cancelled, the two went through a major personal crisis. They had watched virtually every show and memorised every buzzer and bell pressed and remembered the colour patterns of the participants and host. Suddenly one of the things that endowed their life with some sense and order was lost. It was made worse when the host they obsessed about, Dick Clark – 'who we call our personal saviour' – suffered a stroke a few years later. The twins actually got to meet him before and after his stroke and he sends them Christmas cards.

They had a troubled early life, with bad squints, thick glasses and odd behaviour that they and their family were bullied and teased for. Their mother suffered from depression, and on one sad occasion tried to kill them using the gas cooker. She was stopped at the last minute by their father. Their parents, lacking a clear diagnosis, just felt they had odd children and kept them sheltered from the world, mainly from shame and embarrassment.

This went on until, when they were 27 years old, the family with the twins and their younger sister Jane (who had no symptoms) moved from New Jersey to sunny Florida. Their life changed dramatically when both their parents died within four months of arriving in Tampa. Luckily their younger sister was fond of them, and although she had a family of her own, continued to look after them in her house, allowing them much more freedom. Tragically, after only a few years she too suddenly died of a heart attack and they were on their own, confused and lost.

They moved back to New Jersey with their older brother and his family as a temporary measure, but they were having major problems coping, and a care home was the next option. As they were discussing all the deaths in their family and possibly even one day that of Dick Clark, they said: 'We would like to be both buried in a casket surrounded by all the pictures, colour charts and Dick Clark memorabilia.' The thought seemed to give them great happiness.

Dr Darold Treffert, a world expert in these syndromes, examined the twins a few years ago and confirmed that Flo and Kay had the savant syndrome, of which he knew of only 100 genuine cases and a handful of females. He speculated: 'Until we can account for the savant, we can't account for brain function overall. Until we can explain the savant, we really can't explain ourselves.'[23]

Autism, along with its 'milder' cousin Asperger's syndrome (grouped together with non-specific autism), is now best referred to as a group of disorders called Autism Spectrum Disorder (ASD), which can affect up to 1 in 100 of us. Until the 1990s it was viewed as a very rare childhood affliction that was caused by some combination of birth trauma, infections, child abuse, or poor parenting. The term 'refrigerator mother' was used to describe the common psychologist view that unemotional mothers caused the autistic symptoms of their infants and this was a child's self-defence mechanism.[24] In many Asian countries these ideas are still in vogue today, so that the commonness of ASD is largely denied and mothers blamed.[25]

Twin studies in the 1990s changed these perceptions dramatically when they clearly showed one of the strongest genetic influences of all diseases. The original estimates (based on small numbers) show autism to have a very high heritability of around 75–90 per cent, putting the refrigerator-mother theory firmly in its place. More recently, the largest study to date in 192 twin pairs with either autism or Asperger's has suggested that, although still clearly heritable, this figure is likely lower – closer to 40 per cent.[26] High rates of shared autism for identical twins of 60 per cent were found. Surprisingly, the study also found fairly high rates of sharing in non-identical twins, suggesting for the first time that there might be a shared environmental or womb factor, as studies have found much lower (10 to 20 per cent) sharing rates among brothers or sisters of children with autism.

Until recently reports of genes and ASD have been unconvincing, as studies have been small and speculative. However the use of genome-wide methods and large numbers (over 2,000 cases) has eventually produced results. A number of key genes are involved either with altered variations or containing duplications (called CNVs) that alter their function.[27] Although still of small effect, they highlighted certain key brain pathways, particularly genes (cadherin) which encode molecules that allow neurones to interact, as well as a pathway (ubiquitin) which modifies proteins that influence neuronal connections and plasticity, or rarer mutations in a gene altering neural connections that also leads to mental retardation.[28] These gene networks are involved in neuronal function, sensitivity of the synaptic nerve connections, as well as plasticity – the ability of cells to alter function.[29] Some of the changes to these pathways are even reversible in animal models. Overall the gene pathways tell us that minor alterations in neuronal development are crucial in the growing brain. But how could this cause the very special characteristic symptoms of autism?

Cracked mirrors

One of the latest and most elegant theories to explain the lack of empathy and ability to gauge emotions that is seen in autism comes from the eminent neurologist V. S. Ramachandran. Originally an Indian Tamil, he now works in San Diego. His idea derives from the recent discovery that the brain contains groups of highly specialised nerve cells called mirror neurones. These function as a virtual-reality network and are key to the way the brain develops by mimicry.

These mirror neurones can imagine actions and fire off when looking at someone else performing a precise action. The fundamental characteristic of these neurones was that they discharged both when the subject himself performed a certain motor act (e.g., grasping an object) and when the brain observed another individual performing the same act, thus internally rehearsing that act by mirroring the body's own brain response in the same areas. Originally these neurones were shown to be firing in the prefrontal cortex of monkeys by Italian Giacomo Rizzolatti in 1992. The field was slow to accept their importance until the same neurones were shown in humans.[30] Ramachandran and others found that one of the specific brain waves (called mu waves) of a normal person will be suppressed when watching someone else performing an action or expressing an emotion. Autistics however produce no such suppression of these brain waves, or any signs of emotion as tested by skin blood flow or functional MRI brain scans.[31] This suggests that mirror-neurone deficiency (or broken mirror syndrome as Ramachandran called it) is due to disorganised development of the brain, occurring at an early stage.[32] This coincides nicely with the evidence coming from the recent genetic studies.

The fact that most of us have brains that are so full of these mirrors suggests that many of our learning patterns and behaviours revolve around mimicry of others. This gives us ability both to read the emotions and intentions of other humans and to reflect on our own, which is perhaps the basis of human consciousness. We all know about how we mimic other people's body postures subconsciously, depending on how much we like or fear them. Many salesmen now go on courses as part of their training so that they can improve their mimicry skills and sell us even more double-glazing. Autistics, lacking the mirror neurones, don't have this ability to mimic body postures and are generally poor salesmen. In his latest book Ramachandran suggests that the great leap forward in human evolution happened around 60 to 70,000 years ago, after the brain had been at its current size for over 200,000 years without much happening. At that point: 'By hyper-developing the mirror-neurone system, evolution in effect turned culture into the new genome. Armed with culture humans could rapidly adapt to new hostile environments . . .' He suggests more speculatively that this change could also have been the origin of human language.[33]

Ramachandran and others have successfully treated many patients with phantom pain from amputated limbs with simple mirror therapy, enabling the brain to be tricked into rewiring internal pain signals possibly by increasing their mirror neurones. Work has started in autistic patients to see if their few existing mirror neurones can be increased and activated using synchronised dance therapy which trains the brain to mimic others.[34] Currently around 20 per cent of neurones are believed to have this mirror function in humans and are concentrated in a unique and large area, the inferior parietal lobe, at the crossroads of visual, touch and hearing centres of the brain.

The recent optimism for treatment of autism comes from the discovery that the brain is much more flexible than previously believed. Just as we saw with the black-cab drivers whose brain areas altered at the expense of other areas, specialised areas can alter function and cells can alter specialisation through epigenetic change. After strokes or injuries, different brain areas can also take over moving different bits of the body, and visual areas in blind people can convert to hearing or smell. It is likely that the savant skills shown by Flo and Kay operate because other parts of their brain are being underused or the distracting signals flooding in from our normal senses are suppressed, so that they can focus exclusively on remembering trivial details or amazing mathematics. Other savants who may possess, say, great artistic skills when young tend to lose them when their autism improves and presumably they start to use other bits of the brain.

Nerds, hormones and silicon valley

Margaret was looking forward to having twins and had an uneventful pregnancy until the end, when she developed the common complication of high blood pressure (pre-eclampsia) and had to rest. She eventually delivered identical twin boys at 40 weeks in February 1973. Kevin came out first, Shaun ten minutes later, and both looked reasonably healthy.

After a few days, Margaret's happiness dissolved, to be

replaced by anxiety. Shaun was not well and having problems feeding. He had also had a few very frightening grand mal epileptic seizures. The doctors didn't know the reasons but advised keeping him in hospital for a few weeks until he stabilised. Six weeks later, after reuniting them at home, his parents Margaret and Patrick both noticed that Shaun wasn't right, crying very loudly and strangely, whereas Kevin was quiet and placid.

The parents noticed that although the twins looked similar, there were clear differences in how they responded to their environments. By the age of two it was clear there was a permanent problem with Shaun. He couldn't sit or hold things or respond to his name. He wouldn't play or talk with his siblings or use any toys except his unusual favourite objects: plastic bottles filled with water, which he played with for hours. He was fascinated by particular music and records and the only word he spoke for many years was 'Elvis', relating to his favourite record. At the age of six he was given the diagnosis of 'learning disability' and then sent to a series of special schools. Only when he was 18 did his doctors use the label of autism.

His parents were stressed by Shaun, who took up most of their time and energy for the first few years, so they were relieved to see that their older son Mark and Kevin appeared normal, healthy, and played together. In retrospect all was not perfect with Kevin. After the age of five or six, 'He didn't use to look at people directly or greet them in the street, and if asked a question would reply in monosyllables,' remembers Margaret. 'He was also from an early age obsessed with tidiness and washing his hands many times each day.' This was not seen as a problem in a busy family, but thereafter Kevin's childhood, and school in particular, was not easy. He was bullied regularly, often because he wouldn't play sports and would annoy other kids by turning away when a football was kicked at him. In one strange episode he was being picked on at school by a boy. When his dad Patrick went to see the parents he found out the 'bullying' kid was five years younger and half his size.

Kevin often got into trouble, usually because he misunderstood teachers and classmates. 'From an early age I was obsessed with the TV science fiction programme *Space: 1999* [a UK equivalent of *Star Trek*] and the lead female alien "Maya", who could metamorphose, and wore out videotapes watching repeats. I was also fascinated by plane crashes and disasters and knew (and still know) the words to every Abba song'.

He gladly finished school and got a job selling TVs and electronics, and a few years later he bought his own corner shop. He was doing well until one day two masked men robbed him at knife point. This left him very traumatised and he closed the shop. He became depressed for several years and got other jobs, with relative success despite bullying by managers. After fighting with various GPs over his diagnosis and treatment, he consulted an autism support group and was finally referred to Professor Baron-Cohen in Cambridge, where he was formally diagnosed as having Asperger's syndrome. For the last eight years Kevin (now 38) has coped well with his problems with the help of cognitive behavioural therapy. He has dedicated himself to raising money for autism spectrum disorders (ASDs), formed a local support group in Staffordshire, and wrote up his story to help finance it.[35] 'My diagnosis was the turning point in my life. I am a positive and mentally stronger person. I am confident enough to take the stage. I may be nervous – but I have to do this.'

Shaun and Kevin's story is not an uncommon one among twins. They grew up at a time in the 1970s when the diagnosis was rare amid widespread ignorance. They also show clearly that their identical genes are not the only factors and can produce a wide variation in outcomes: from an intellectually impaired and totally dependent brother to another who is a high-functioning ASD with independence and a successful life. While it's possible that some birth-trauma difference between the twins accounts for these brain differences, it is more likely that these changes in Shaun occurred before birth, causing brain damage and making the brain liable to seizures, which in turn can make things worse. The genes may have given then both a mild susceptibility, but something else in the environment had accelerated the process and altered the way the genes functioned at a crucial stage.

Of all the changes in human diseases, recent changes in autism rates have been the most dramatic, puzzling and controversial. Over the last three decades autism has increased tenfold, from 4 per 10,000 children in the 1960s to around 40 per 10,000 today in the US. More than 3,000 new cases of autism were reported in California in 2006, compared with only 205 in 1990.[36] The numbers have continued to rise since then and are mirrored in most but not all countries. Although some of this increase (maybe half) is due to changes in diagnosis and medical attitudes, there appear to be real increases of 2–3 per cent per year.[37] This rapid increase and the odd twin results can't be explained by traditional genetics, so what is going on?

Previous research has implicated in a few cases anti-epileptic drugs, thalidomide and measles infections in mothers in early pregnancy, as well as a link with cerebral palsy. Up to a third of autistic children have had some form of temporal lobe epilepsy when young, which could certainly also upset brain development. Researchers recently found a higher risk of autism among children born to mothers who took SSRI antidepressants (like Prozac or Zoloft) during the year before birth, particularly in the first three months of pregnancy.[38] But none of these exposures have increased dramatically enough recently to explain the present trends.

One popular theory was that the rises were related to increased childhood vaccinations since the 1960s. In 1998 *The Lancet* published a UK study implicating a triggering role of vaccination with the triple combined MMR (measles, mumps, rubella) vaccine. The paper, and media frenzy promoted by the authors, caused millions of parents worldwide to boycott the vaccine and led to a major recurrence of measles and a number of deaths. The paper was later retracted, due initially to the main author, gastroenterology surgeon Andrew Wakefield, having a major conflict of interest.[39] He had been working as a paid consultant for the law firm trying to sue the vaccine manufacturers via a class action, and had acquired his 12 classical autism cases mainly through this route. Much larger population studies have since shown no association between use of the vaccine and ASD. Wakefield was subsequently investigated by the GMC and stuck off, with a suggestion of fraud. It appears on closer scrutiny by investigative journalists like Brian Deer of the *Sunday Times* that at least five of the 12 cases already had developmental problems before the vaccine.

This story was a tragedy all round. Not only for the normal kids who developed some severe complications of measles like blindness and mental problems, but also for the autistic families who had clung to the idea that this was the cause of the problem – which made it much easier to accept. Unfortunately it was just a coincidence: autistic symptoms happen to become apparent at much the same time that vaccinations are often given. But after this major red herring, are we any closer to finding a cause of the increase?

Another current theory proposes that the increased mating of nerdy men and brainy women explains reported high clusters of disease in areas of the US (Silicon Valley and Cambridge, Mass.) and UK (Cambridge) with hi-tech industries and computing. This is hard to prove and prone to bias, although it would go along with the extreme male brain theory originally proposed by Simon Baron-Cohen.[40] The theory goes that ASD sufferers are at one end of a spectrum of male–female differences, whereby typical male (highly technical or systematised) brain activity is favoured over emotional or empathetic activity (typically female). As males with high systematised intelligence (sometimes called geeks) are now finding great financial success in biotech, computing, electronics and gaming, they have apparently become more attractive to some females. Indeed, if you are interested there are growing numbers of websites dedicated to women seeking nerdy husbands (such as www.Nerd Passions, Geek2Geek, IQcuties, Sweet on Geeks etc.).

The theory would suggest that whereas in the past these men would have been monks sitting alone on high wooden stools copying Latin texts and performing amazing calligraphy, now they are back in the gene pool. These matings are more likely to produce high-IQ males with ultra-male-orientated brains and at risk of ASD. The nerd mating theory is pretty much impossible to prove or disprove, although it might explain a few cases.

A better bet for a broader explanation of the increase in ASD is hormones. The extreme-male-brain theory has fostered ideas that levels of sex hormones in the amniotic sac of the womb –in particular testosterone, produced by both the fetus and mother – are quite variable and could influence the brain. A study in Cambridge has been following up 635 children for ten years having obtained amniotic fluid samples at birth. So far they have found good correlations between levels of high testosterone and mild features of ASD such as avoiding eye contact, reduced empathy and reduced communication skills. Contrary to the views of half the population, men actually have larger brains than women. Although size isn't everything, men also have much larger amygdalas – paradoxically a key area in emotion and empathy. ASD sufferers have larger-than-average brains and amygdalas – again supporting the idea that ASD males have excessive male characteristics and that sex hormones play a role. It is also possible that the X chromosome has a protective role, as females have two copies of all these genes and males only one, allowing females more leeway if one copy of their genes gets modified epigenetically or rare mutations occur.[41]

There is now good evidence of an epigenetic influence in ASD.[42] First, two common single gene imprinting disorders (which you will remember are extreme forms of epigenetics by which one copy of the gene is totally switched off) – fragile X syndrome and Rett syndrome – are both associated with autism and ASD. Second, a few ASD patients have gene duplications in a key area of chromosome 15 responsible for other similar disorders (Prader-Willi syndrome and Angelman syndrome). Finally, a small study of three discordant identical twin pairs with autism found small methylation differences in key candidate genes for ASD, and methylated suppression of a gene that reduces a protein in autistic brains.[43]

But if epigenetics is a mechanism by which genes could be modified, what unexplored external factors could be responsible? Some studies have implicated older parental and grandparental age at conception.[44] As women (and men) are gradually increasing in age at the time of having children, this could cause genetic and epigenetic instability in sperm and eggs.[45]

Another study looked at the month of conception. The study included more than 6.6 million children born between 1990 and 2002 in California, of whom 19,238 were diagnosed with autism before the age of six. Children conceived in the winter season (December, January or February) had a 6 per cent greater risk compared with those conceived in July. Greater numbers of winter infections or vitamin deficiencies are obvious suspects.[46] Infections like viruses have long been implicated in ASD, but the link has never been proven. Moreover childhood infections have actually reduced in recent years.

Vitamin levels before birth may have changed. Some nutritionists believe that the nutrition content of most fruit and vegetables we eat has decreased over the last 50 years. If true, this could account for a number of recent trends. Twenty-seven vegetables, 20 fruits and ten meats were compared over 50 years by careful surveys in the UK from the MRC and Ministry of Agriculture. They showed average falls of 48 per cent in calcium and 27 per cent in iron, with similar or greater losses for other minerals like copper or magnesium.[47]

A recent case-control study (not a proper trial) of 288 autistic children in California shows that use of prenatal vitamins around the time of conception was associated with a reduction in the risk of having children with autism.[48] They found an overall 40 per cent reduced risk if mothers took vitamins. Some influence of these vitamins on gene methylation was also shown. A few women not taking any vitamins who also had certain gene variants influencing levels of methylation had a risk of autistic children twice to seven times as high.

There are probably more theories for the cause of autism than for any other disease, each with its own website and support group. Those mentioned here are just the mainstream ones. Another has appeared recently suggesting that paracetamol (acetaminophen) could be the deadly cause, as use has increased since the 1980s. This sounds crazy, but supporting it is the strange phenomenon that when autistics develop a fever their autistic symptoms often lessen or disappear, implicating the fever centre of the brain (the locus coeruleus). Artificially inducing fever has now been proposed as a therapy, but as you can imagine sounds medieval and will be tricky in practice.[49]

Plastics, poisons and pregnancy

There is a nearly limitless number of theories and conflicting opinions in the field. Some sceptics are not yet convinced there is even a real increase in cases of autism. However, if we accept the majority opinion that it is increasing and agree that males are four to five times more at risk, then the clear link must be a hormonal influence of some sort on the developing brain. This would have to have changed rapidly in the last 30 years.

We have discussed vitamins, which might partly contribute, but what about chemicals? We talked already about some – called hormone disruptors – that alter hormones epigenetically. These include phyto-oestrogens from plants that can influence breast-cancer risk, cadmium in tobacco and low levels of arsenic. But there are others, that if you don't live in Canada you may be less aware of. A year after Walter Brooke's famous quote to Dustin Hoffman discussing his future in *The Graduate* – 'I have only one word to say: "plastics"' – the world changed, and became full of them. One plastic component under the microscope is called BPA or bisphenol A, found in most plastics used today since the 1950s,[50] manufactured in vast quantities in the US and other large countries. It has mainly oestrogen-like properties and has been tested quite extensively over the last few years in rodents in realistic (i.e. not ridiculously high) doses.

Like other hormone disruptors, it has some cancer-inducing effects in hormone-related cancers, as in the prostate and breast. However, its role in the brain is most interesting. BPA when given as a supplement to our old friends the Agouti mice will alter their coat colour by epigenetically removing the blocking methylation signals on the gene (and therefore activating it), an effect that can then be reversed by giving methyl or folate vitamins.[51] More worryingly, mothers' exposure to BPA has also been shown to alter methylation in the fetal mouse forebrain. This produces changes in the neuronal synaptic connections, and in brain development as well as behaviour in the offspring.[52] One of the clearest effects is increasing anxiety in pups and altering the mother–infant behaviours, making mother and pup more introverted and the mothers less caring. One study was performed on pregnant monkeys given BPA for the last 28 days.[53] Infants of these BPA mothers were physically similar, but less attached physically and visually to their mothers. Also relevant to autism, BPA in the low doses that might be found in humans has been shown recently in male rodents to reduce two different learning skills in a maze test. It may do this by altering the hormone receptors in the hippocampus.[54]

Another study looked at mice offspring of mothers fed BPA in pregnancy. These were special male deer mice who in the wild need great visual–spatial skills to find and mate with females across long distances.[55] The BPA mice looked identical, but were a bit geeky underneath and had poor female-hunting skills compared with normals. The females sensed this somehow, and even when the BPA mice did eventually locate them they found them less attractive. As well as giving BPA in pregnancy, studies have also looked at BPA in the male line where it affects sperm and promotes testicular disease and can be passed on epigenetically across at least three generations in rats.[56] These are just the results for BPA. Similar results occur with other chemicals such as those found in flame retardants called BDE-4 – shown to alter rodent learning and induce behaviour inactivation[57] – or with vinclozolin, the chemical used in many common fungicides, which reduced male fertility even in the fifth generation.[58]

So should we all be worried by plastic? After all, it is now pretty hard to avoid. Try it yourself – BPA is the type 3 and type 7 marked on recycling codes. Over 2 million tonnes are produced every year and it is the major hormone that leaks into our rivers and water supplies. When tested 95 per cent of us have detectable levels in our urine.

Although the risks in animals have been known for over five years, governments have been slow to react. Canada has as of September 2010 banned BPA as a toxic substance, and the EU has recently banned its use in baby bottles. In the US, despite a worrying report to the FDA in 2010, no action has yet been taken. Those with faith in government will rest assured because the director of the US Food Standards Agency and the UK Medical Research Council rated it 'quite safe in humans' as doses are low and it was all excreted. But it's not just baby bottles: BPA is everywhere – sports equipment, CDs, DVDs, medical devices and tubes, recycled paper, dental fillings, even the inside of most canned drinks, and on plastics in contact with food. Worryingly, the highest levels (400-fold) so far recorded are in babies in intensive care units, surrounded by plastics, where it can be stored in fat tissue for years.[59]

In humans observational cross-sectional studies of BPA levels in the urine in 1,445 Americans measured at different time points from the NHANES study linked them to an increased risk of breast and heart disease,[60] diabetes, altered liver function and many other slight changes in blood tests.[61] The chemical triclosan, found in deodorants, is similar and in humans is associated with altered immune function.[62] Although these kinds of observational studies can be misleading, they are worrying.[63] Although the levels detected are within the safe limits suggested by US and UK governments, doses lower than that have been shown to affect rodent behaviours. But let's not forget the other 104 known hormone disruptors, including pesticides, herbicides, fungicides, hydrocarbons and flame retardants, that could be having similar or unknown effects on us or our grand-children. Food for thought as you swig from your plastic bottle of ultra-healthy mineral water – which when tested has oestrogenic properties in 61 per cent of samples, unlike glass bottles or tap water. So plastic bottles may be one reason why women's breasts (and men's) are getting larger.

10

'THE GAY GENE'

Sex, hormones and the brain

All this talk of deadly chemicals in our atmosphere may have made you wonder what other changes they might cause. Some experts believe our sperm count and quality may have halved over 50 years – matching an increase in testicular cancer.[1] Although environmental pollutants have been blamed, there is considerable controversy and conflicting data about the scale of this. In China a recent study showed low sperm counts in bisphenol (BPA) workers in plastic factories.[2] Other studies have suggested that our rapidly rising obesity levels are the main cause of lower sperm count. While meta-analyses have not shown a direct effect of obesity, they do show that obesity is somehow associated with lower testosterone levels.

Could researchers be looking in the wrong place at the wrong time? When the mothers are studied, two important factors show a clear influence on lower sperm counts in their sons: obesity and smoking. These effects are actually larger in mothers than when looking directly at the sons.[3] In particular maternal smoking seems to have a disruptive effect on the hormone levels in the fetus.[4]

Most of the evidence in humans is still indirect, but a key point is how low the doses of these toxins and hormones need to be in humans to produce an epigenetic change in the brain. I believe part of our individuality could come from these subtle influences causing changes in our genes and hormones at critical points in our brain development. Quite how subtle these changes are is clearly illustrated in animals. A small chill in the air for a caterpillar causes the female butterfly (like the squinting bush brown variety) to change from a shy demure creature and act as a predatory female. They have brighter wings and actively chase unfortunate males until they are relieved of their sperm and other valuable nutrients.[5] In many reptiles and some fish a change of one degree in water temperature (or an ultra-diluted sample of bisphenol chemical) can alter the sex-determining genes and change a male fish into a female.[6] Some of these temperature-dependent sex changes can be reversed by addition of hormones or hormone blockers,[7] demonstrating the close relationship. A few bees fed royal jelly become super-fertile queen bees and the rest, fed on a weaker brew, become sterile short-lived workers. But other gender-determining signals in humans are even more mysterious – like the fact that the normal ratio of male to female births (currently an excess of around 5 per cent) varies over time – and in the last 50 years has been showing a male deficit in the US and Northern Europe.[8]

On the first day of life a baby girl reacts differently to a boy in the way she looks at faces or objects, and after only three months, clear toy preferences can be observed of trucks versus dolls. These reactions are also observed in monkeys, so they cannot be due just to exposure to the Barbie doll adverts that have programmed girls to love pink for only the last 50 years. Sex determination and development of ovaries or testicles occurs early in fetal life at around 6–12 weeks, and sets the crucial gender environment for the rest of brain development.[9] The default gender program for humans is the female model (unlike Adam – the Garden of Eden version). If during these crucial weeks the gender-determining (SRY) gene on the Y chromosome produces adequate testosterone levels, changes will allow the development of a male. If the testosterone signal is below the threshold, a normal female will continue. After this crucial crossroads the brain starts to develop and organise itself with gender in mind, although many of the gender-specific pathways may not be switched on until the genes are activated by the tidal wave of hormone signals at puberty. Our brains and hormones don't always work in ways we expect.

Gender secrets and lies

Chanelle and Gabrielle Pickett were good-looking identical African-American twins. When they appeared in 1993 on the American *Jenny Jones* TV chat show they looked attractive and very alike, with straight hair to their shoulders and light pink dresses showing off their good legs and figures. They got a mixed reception when they walked on to the show but were relaxed and articulate. The reason they had been invited – and for the audience's odd reaction – was that ten years before, they had been cute twin boys.

At the age of around seven, Chanelle and Gabrielle both realised they were very different to other boys. They started dressing up in Mum's clothes together, stealing make-up and secretly putting on lipstick and eyeliner. At first they did this in secret, then they got bolder, until by the age of 13 they started going out together wearing dresses. One day soon after, their mum found them at a downtown Gay and Lesbian Youth Center and she flipped.

'We told her we knew we were different and had some problems we needed time to sort out. She was a very different person after that,' said Chanelle. In fact she refused to have the 14-year-old kids back, and they ended up being fostered in another part of the country. Here in anonymity they dressed as girls permanently and had a great time in the local high school, where they were sociable and well accepted. That was until the headmaster told his staff 'confidentially' their secret and word rapidly leaked out to the pupils who had once been their friends. School became a nightmare: they were bullied and harassed. 'It was horrible – we were suddenly treated like a freak show,' recalled Gabrielle.

They didn't stay in school long. They both worked hard with short-term jobs to save up for a male-to-female sex-change operation, where the testicles are removed and the penis inverted to form a vagina. With the extra confidence this gave them and with the help of regular oestrogens to stop hair growth and promote breast tissue, they were hired for better jobs. But trouble was never far behind. Although generally popular, one of them would find themselves being picked on or bullied by a colleague or boss, and they would have to move on. Although they held some part-time jobs, over time they drifted into the underground world of the transsexual sex scene, where straight male punters would pay to have sex with them. Incidentally, many straight men are turned on by transsexuals, and websites with T-girls are the fourth-most-popular Internet sex category.[10]

One night Chanelle was in one such (TS/TV) club called Playland in Boston, when in walked a regular, a computer programmer called William Palmer. He took Chanelle back to her flat, offered her some cocaine, and tried unsuccessfully to have a threesome with her twin. Although apparently no money was involved, they then went back to his flat. At 5 a.m. Palmer shouted from his bedroom: 'This is not happening, keep quiet or you'll wake the whole neighbourhood.' The shouts turned to screams and pounding on the wall for minutes, which slowly diminished and then suddenly stopped. Palmer's flatmates were worried and tried to open the door, which was blocked. 'I've got a crazy bitch in here, but I have it under control,' he replied, and they went back to sleep. The first thing Palmer did in the morning was to phone his lawyer.

Chanelle, who was only 23, never awoke – she had been strangled for eight minutes, her face badly beaten and had bedding stuffed in her mouth. At the trial in 1995 Palmer claimed he hit her once then went to sleep. His slick defence attorney said he hadn't known that Chanelle was a transsexual and this must have provoked the argument. The jury acquitted him of murder and he was given a two-year sentence for the minor crime of assault and battery. Most commentators agreed that the result would have been very different if she hadn't been transsexual, he hadn't been white, and she hadn't been black. Her sister Gabrielle wasn't any more fortunate. She too was murdered in mysterious circumstances eight years later – they never found her killer.

What makes transsexuals like the Picketts go to these extremes? They changed their lives, underwent traumatic and costly operations, took dangerous hormones, and risked becoming social outcasts. In the US, there have been over 30,000 operations performed since the first in 1930. This is dwarfed by over 100,000 in an unlikely country – Iran – where the government pays for the operation as a cure for homosexuality.[11] Being transsexual describes individuals who identify themselves with a gender different to their biological sex. People who feel like this, but don't appear in sex-change statistics, are more common than you might think, and their numbers are likely rising. In the past it was thought to involve just 1 in 100,000 biological men, but current estimates are that around one in 5,000 men feel they were born in the wrong body.[12] Inversely, female-to-male transsexuals account for about a third of cases. Some of these women have vivid phantom penises from early childhood and some even report phantom erections.

Men can have feelings that their penis is a foreign object that doesn't belong to them and that they want removed. This may be a way of the brain attempting to resolve inner conflicts between the concepts of gender identity and self-image. During our fetal growth several processes of sexual development occur in parallel: the sexual organs, sexual identity, sexual attraction and sexual body image. Normally they are in tune with each other, but as we see they can often drift down different paths – particularly under the influence of hormones. Any mismatch between them can lead to physical or psychological gender identity problems that can make terms like sex and gender sometimes difficult to define, even for experts. This is exemplified by externally female athletes like the South African runner Caster Semenya, who has three times the normal testosterone levels of a female. She is reported by the Australian press to have external female genitalia and internal testes and no womb. She has been brought up as female and has been passed by the International Athletics Association as female for competitions.

Testosterone, finger length and the brain

Although the Picketts were identical twins, their case is relatively unusual, and most transsexual twins are discordant for gender identity – suggesting that subtle hormonal or epigenetic changes occur more to one twin than the other.[13] There is also some evidence of an overlap between transsexual behaviour and autism/ASD, supporting a role of the 'masculinised brain' in at least some female-to-male transsexuals. The link to autism also supports a key role of differences in brain development affecting the areas involved in recognition and self – linked to gender identity, and another link between genes, hormones and the mirror neurones.[14]

As fetal testosterone or oestrogen cannot be safely or ethically measured directly, we have to look at non-invasive indirect methods of assessing it. Our adult bodies retain a trace like a fossil of the chemical and hormones we were once exposed to in the womb. Take a look at your right hand flat in front of you. Ignoring the thumb, you can see that the four fingers have different relative heights. Focus on the difference in length between the second digit (index) and the fourth (ring) finger. If you are female your second should on average be equal to or slightly longer than your fourth finger. If you are male your second digit is usually shorter than your fourth digit. Differences in this ratio could be a marker of your personality, aggression, criminality, sexuality and sporting prowess. These relative measures are known as 2D:4D digit ratios and reflect the relative levels of the male and female sex hormones in the womb. Broadly speaking – although there is much individual variation – the lower the 2D:4D ratio, the more of a traditional male pattern, and the greater the ratio (i.e. longer index finger), the more of a feminine pattern is represented.[15]

Quite why our fingers vary in length according to gender and sex hormones is unclear, as having a long ring digit as a male or a long index as a female doesn't seem to confer any particular physical advantage that I can identify. But they do seem to be rough guides to how differently our bodies may have developed – in particular the brain and maybe the genitalia. Other Sherlock Holmes-type clues as to how androgenised or not you are include having a hairy ring finger, an early age of puberty, the ability to hear fine sounds from the inner ear (oto-acoustic emissions), and something you may not have measured recently: the distance between your anus and your urethra or penis.[16]

Variations around these measures are associated in rodents and humans with different sex-related traits such as fertility and sexual preferences.[17] These changes occur early on in our development, shortly after the sex-determining genes are turned on, at around 12 weeks gestation. This is a time when this process is very sensitive to any changes and when natural and non-natural factors play a role epigenetically.

John Manning of Swansea University was the first to speculate on this magic finger ratio in 1998,[18] although it had been noticed as an oddity since the nineteenth century.[19] Nowadays it is nearly a medical speciality in its own right, with over 500 published papers on the area.[20] Some of the most consistent studies showed that the smaller the ratio in male or females, the higher their level of sporting achievement.[21] Other studies, not as consistently, have shown that self-reported masculine traits (deep voice, upper body strength) are also associated with testosterone levels, as well as with tendencies towards crime, facial attractiveness to women,[22] and skills at computer programming – suggesting they are all under some sex-hormone control.

Although the size of the penis has in folklore and jokes often been associated with the size of the hands and feet, these chestnuts have been discredited. However, a recent study of Asian fingers has shown a reasonable correlation between having a more masculine 2D:4D ratio and a larger penis. The clever Koreans did this by identifying patients due for prostate surgery: when anaesthetised the diligent researchers stretched the flaccid organ (of the volunteers) and measured it, comparing it with the finger ratio. The stretched organ apparently gives a better indication of the erect size than the floppy one. While the press presented this as a major breakthrough – and made a handshake with Asians a potential party trick – the stats behind the study showed that as the correlation was quite weak, for any given individual its prediction was pretty useless (explaining less than 5 per cent of the variability in member size) and likely to lead to disappointment.[23]

A couple of genetic studies have shown that genes are at least partly responsible for variations in the magic ratio and possibly indirectly fetal testosterone. Our own twin research as part of a large collaborative study showed (as usual) a heritable component.[24] A key gene associated with puberty development was convincingly associated with the ratio, but unfortunately in the opposite direction to that predicted.[25] They did not find any other genes, but the study was relatively small and certainly couldn't rule out the effect of genes on the testosterone pathway.

Other relics of our past fetal androgens have not been studied as much, but researchers in Canada painstakingly counted the hairs on the back of the middle section of the ring finger in women.[26] They found that having more hairs than the average seven (and a more masculinised profile) protected women from some of the hormonal side effects of the oral contraceptive pill. For strange reasons, if you look at the same middle section of your index finger you will find many fewer hairs (assuming you don't shave or pluck your hands). In fact, the average was less than one hair. This bizarre biological quirk suggests a straightforward relationship between circulating testosterone and how masculine a male will be (in terms of sexuality and behaviour and body size) – with finger ratios and hairy fingers a marker of this. Yet, as usual, relationships can be deceptive.

Male homosexuals are on average less keen on participating in and less gifted at some sports than heterosexual males.[27] Also, aggressive male criminals have been shown on average to have larger penises, so male homosexuals might be expected in theory to have smaller genitalia than heterosexuals. However, studies actually show the opposite. Gays have been reported to be actually better endowed on average than heterosexuals.[28] This (like most penis surveys) is usually based on self-report and so related to (and biased by) body image and confidence, although it could be argued that homosexuals may be more accurate, as they have considerably more experience in comparing themselves than heterosexuals. Homosexuals also have a strong male-type sex drive, although directed at other males. Lesbians in contrast (but not bisexual females) tend to have more typical female sex drives despite having longer ring fingers and being better at sports. So changes in testosterone levels probably have different and sometimes opposing effects on the developing brain, depending on the absolute levels, relative ratio to oestrogens, and also the precise timing and duration of the changes.

And let's not forget the genetic background. While most agree that it has a role, the evidence for fetal testosterone accounting for most of the sexuality differences we are discussing is too simplistic and overstated.[29] The lesson from biology is that the real explanation is nearly always much more complex.

Gay genes and randy aunts

Nigel knew he was different when he recalled at about the age of 13 'not feeling the same about girls as my friends did. I started to have erotic dreams about men and realised in retrospect that I never had any dreams about girls.' No one in his family was gay, he found these feelings frustrating, and he became an angry teenager for the next three years. At a party, aged 16, he went for a walk with a girl and kissed her but remembers 'it did nothing for me'. At age 17 in his last year at school he was teased by other kids for having an 'inseparable friend – Simon' and was jokingly called his boyfriend. They laughed it off, but Simon had actually been with other boys. Secretly the two got to know each other intimately and were soon engaging in masturbation and oral sex, although Simon also had a girlfriend he was cheating on. Nick realised at this point that he was definitely gay. Soon after he left school and home to be a chef, he entered the gay community and didn't look back. He told his family soon after, including his twin brother Mark, who remembered: 'I wasn't that shocked to be honest. I suspected it at his last year at school – but not before.'

Mark is a nurse and like his brother hated sports – which in New Zealand was not the popular choice. They both preferred arts and music. 'I was a late starter with girls and waited until my twenties to have sex and a serious girlfriend. I've never had any dreams about men, but I like and prefer the company of girls and gay men. Nigel often used to take me with him on the gay scene and I used to get offers.' One night when Mark was 22, a particularly good-looking and persistent man chatted him up at the bar for hours and bought three bottles of champagne. Eventually he agreed to go back to his place, where they kissed and got undressed. 'Suddenly naked in bed with this man I realised I wasn't interested. I got up, apologised and left. I realised Nigel and I are very similar in nearly all respects, but in choice of partners I just wasn't like my brother'. He has now been happily married for five years and has a daughter.

Nigel is 8 kg lighter than his brother, he runs several pubs and has also had a steady partner for the last five years. Their parents are relaxed about him being gay. Their older brother is straight and their mum comes from a large family with four other sisters and 11 other cousins but – as far as they know – no other gay kids. Neither twin is at all interested in sport or does any exercise. They have similar digit ratios close to 1.0 (the index and ring finger being closer in size) – higher than most men.

A similar story comes from Daniel and Thomas, retired actors born in a different era, the 1940s, who are also identical twins. They were pretty, handsome boys with very similar personalities, preferring arts to sport, and both matured late in life. Tom was always more interested in girls, while Daniel was more reticent and felt less comfortable at school dances, although the girls seemed to sense his relaxed, indifferent attitude and he was popular.

Tom was the first to have a sexual encounter, a fumble in the back seat of a Hillman Hunter with a 16-year-old would-be seductress. He recalls: 'It was an embarrassing incident I tried hard to forget. I was very anxious and excited and came far too early and it was all over in a moment. I don't think the girl was very happy.' At different universities, they both decided independently that they should properly lose their virginity. Daniel duly found an obliging girl at a freshers' week party, but it ended in disaster, with much fumbling, anxiety, embarrassment and a lack of an erection.

Tom, who was not put off despite his previous back-seat disaster, found a friendly girl, consummated, fell hopelessly in love and didn't look back. Daniel, not keen to repeat his experience, soon met a boy and decided that it was boys that he was aroused by. Sex followed and he too fell in love. 'After that I never really considered women sexually again'. Tom did have some nagging doubts. 'Once aged twenty-two I went home with a good-looking gay man after a drunken actors' party and ended up in his bed. In the 1960s it seemed like everyone was trying it. I was very drunk and to be honest not sure if penetration actually took place. But I then realised that sex with men was not for me and I definitely much preferred women.' Both twins are popular with both gay men and women. They agree that if anything, Tom has a more feminine side to him than Daniel, but neither has since been tempted to swap places.

Human sexual behaviour falls within a wide spectrum from complete heterosexuality to complete homosexuality, with all possible combinations in between, varying in frequency and throughout life. A joke goes in the gay community: 'What's the difference between a straight man and a gay man? – Six pints of beer.' The truth is that for most men it will usually take more than that. The frequency of self-declared homosexuality – defined as some adult sexual contact with a same-sex individual – is around 5 per cent in the UK for males and 2 per cent for females.[30] Rates are reportedly higher in the US and in France, and rise to 16 and 17 per cent if you include any instance of sexual attraction. This appears true across all human populations and across any time studied.

What is interesting about our male twin cases and others I spoke to is the lack of bisexuality in the twins where one was gay. The straight twin in both cases did try switching once – probably consciously or subconsciously to ascertain if they might have some hidden gay tendencies. However, their single gay experience ended up reaffirming their original belief that unlike their twin they were straight. This is despite being very close to their twin in personality and other tastes, which suggests very specific differences in the brain. Most men, once they have had a gay partner or partners for a while, don't revert to preferring women. The number of self-reported bisexual men equally turned on by men and women is actually rarer than we think – although they certainly do exist.[31]

Sexual preference is in part genetic, as shown by several twin- and family-derived heritability estimates ranging from 30 to 50 per cent.[32] Twenty years ago Dean Hamer at the NIH claimed to have found 'the gay gene'.[33] Yet the discovery, although plausibly on the female X chromosome, disappointingly couldn't be replicated.[34] Indeed, as for nearly all complex human characteristics, there is not going to be a single defining 'gay gene', but hundreds of gene variants of small effects influencing susceptibility. Although many homosexuals marry (around 50 per cent in some studies), the average number of kids they conceive is much less than heterosexuals. So according to Darwin these genes should have died out long ago – unless they also have some hidden advantages.

One theory proposed initially by Simon LeVay is that the genes in females from the same gay families increase fertility and fecundity to compensate for the losses in the gay men.[35] There is growing evidence that the wider families of gay men tend to be larger – particularly on the female side.[36] The family's gay genes in the women could increase the strength of their attraction and lust for men (androphilia), giving them more sexual partners. These 'gay aunts' may also have an increased level of fertility, by also having favourable gynaecologic and obstetric factors.

Another theory that has been proposed to explain homosexuality is the 'immunisation of the placenta theory'. Gay men tend on average to come from larger families and often have older brothers. Studies have suggested a doubling of chances of being gay if you have at least two older brothers. The idea was that the first and second brothers' testosterone would interact with the immunology of the mothers via the placenta, so that by the third son, antibodies to either testosterone or the male-determining gene or protein had been formed.[37] A similar relationship has been shown with birth order in male-to-female transsexuals.[38]

I find this idea a bit far-fetched, and apart from the fact that gay men do tend to have older brothers more often than not[39] there is no hard evidence for any immunological differences.[40]

An alternative (and speculative) explanation might be that in families with an over-representation of males, lacking X chromosomes, subtle differences might occur in epigenetic reprogramming of the germ cells so that they retain a trace of some testosterone-altering genes or signals. The same process might explain why some families for many generations can produce only sons or only daughters. There is some supporting evidence for this in studies of rodents.

In many cultures extreme environments can produce temporary changes in sexuality in most people. A good example is men and women in prisons, where some studies suggest that up to 50 per cent of males participate in homosexual activities and 20 per cent report abuse or coercion,[41] although most would not call themselves homosexual – just adapting to circumstances.

In traditionally isolated and religious cultures like Afghanistan, where female sexual repression is the norm, dancing boys (aged ten to 14) are commonly employed for male parties to perform various sexual acts enjoyed by up to half of Pashtun men. This is considered the social norm and not a 'gay' activity, with these unfortunate children seen as available female substitutes. In an unusual boy scout custom the Sambia tribe in an isolated part of the highlands of New Guinea require the boys aged ten to 12 to every week fellate the older boys aged 14 to 16 in the belief that it is good for both groups. It 'apparently' helps the older boys prepare for manhood and having sex with women. Swallowing the semen also 'helps' the younger boys grow into men and become stronger.[42] The Sambia men when older seem normal and are attracted to and marry women. They have reported adult homosexual rates of around 5 per cent – around the same as most other populations.

These unusual groups, and the identical twin pairs I interviewed who shared very similar lifestyles, suggest that the environment while growing up is not critical in permanently changing gender preference. However, the growing brain while in early development is certainly very sensitive to subtle changes, and in different individuals even with the same genes can produce different amounts of masculinity, aggression and sexual preferences, which are not always related.

As we saw from our case histories and confirmed by the larger twin studies, most twins (around 70 per cent) with identical genes do not develop the same homosexual preferences as each other. This shows that genes alone are not enough. There are rare stories of men who become homosexual after strokes or others who change sexual identity several times in a day – a state known as alternating gender incongruity or AGI. These rare examples show the potential plasticity of sexuality.[43] In conversations with gay men about ongoing research into discovering gay genes – now inevitable within a few years – many expressed concern. They were worried that the excuse of genetic causation would be used by the religious right to screen babies and try to convert them by prayer or hormones, or would lead as in India or China to abortions or infanticide of females.

But our studies suggest that even identifying all the possible gene variants will not itself permit the accurate prediction of homosexuality in an individual. So far, religious gay conversion programmes (common in the US and in countries like Malaysia, where 'sissy camps' are compulsory) with bonding, wrestling, prayers and group therapy at $500 per day have not ascertainably succeeded.[44] Once the brain has developed, it seems unlikely that any chemical or hormones acting epigenetically would be effective in childhood. So these fears seem for the moment unfounded. For males, it appears that in most cases early life experiences are not crucial to later preferences, and that once a choice for homosexuality has been made, males rarely revert to preferring females.

Femininity, fluidity and flexibility

Women are quite different though. The fluctuating, fluid sexuality of bisexual women over a lifetime is well documented, showing that female sexual identity is not the rigid construct we imagine.[45] In contrast to the treatment of gay men, societies seem more tolerant: there are few 'butch conversion camps' for females. Gay genes in females are perhaps just genes for flexibility or plasticity that in some circumstances or environments can lead to being gay or bisexual, and in others to being straight – but all in the same life.

Louise was seen as a bit backward by her family. She was painfully shy and more of a loner than her identical twin sister Barbara. She was inclined to be moody and sometimes depressed. 'I lived in a bit of a dream-world I suppose. I always thought my sister was more attractive and successful than me. She was really into boys at age fifteen, so I reluctantly went out with her and other boys because of loyalty and social pressure.'

Louise's first sexual intercourse with a boy occurred when she was 18. 'It was OK, I suppose, but nothing special. Although I was a bit of a tomboy, I never really thought that I was gay, but definitely noticed something about me was different. Although I dated and had sex with about ten to twelve boys, it just never felt really right.' She had her first lesbian relationship when she was 27 and 'At last it did feel right.' She has since had another ten female sexual partners and had her longest relationship – 12 years – with a woman. For the last four years she has lived with an older female partner who used to be married and has grown-up kids.

Her sister Barbara remembers that by the age of 11 'Louise had already started to dress in a more masculine way. By the age of fifteen when I was interested in boys my sister would find excuses to stay home and study. I felt she never respected men and this caused a rift between us. Once, because I know we have the same genes, I was worried whether I too was gay, but when one of Louise's girlfriends tried to sort of seduce me, I realised I wasn't at all interested.'

Barbara has had a total of five sexual partners, has been married with three kids for over 20 years and has always been faithful. Louise, now aged 55, does still sometimes fantasise sexually (and dream) about men. 'It is odd that I do this. These usually involve submissive/dominant situations, like him tying me down or forcing me to have oral sex.' Last time she slept with a man was about six years ago: 'I blame the alcohol – but it was quite fun.' Her ring finger is slightly longer than her index and longer than her sister's, and compared with her sister she was always a bit better at sports.

This story is not untypical of a lesbian woman, who will often have had sex with equal numbers of male and female partners and switch back and forth over a lifetime.[46] Only a small minority remain solely homosexual throughout life.[47] In our recent study of 4,400 UK twins, female homosexuality was around 25–30 per cent heritable,[48] which was slightly lower than for males and similar to the 19 per cent found in a large Swedish study.[49] This suggests a greater relative role of chance or the environment. For lesbian or bisexual women – although research is sparse – no clear environmental components have been consistently identified. The key differentiating factor is that in females as opposed to men there is much more lifelong flexibility and plasticity. The genes for this trait could be providing a slight relative advantage by aiding greater sporting prowess, more masculine thought patterns and the option of finding both males and females potentially attractive – which improves social networking. When women perceive that they are gender-atypical, when mild it may lead to having more sexual partners and offspring, and when severe – which is rarer – it may lead to being a lesbian with no offspring. The twin studies show us that simply having identical genes is not enough to make someone gay. Other subtle environmental differences in hormones or chemicals are needed, acting on our brain neurodevelopment – again probably through epigenetics.

Dean Hamer and his team looked some time ago at differences in mothers of gay men to see which of their two X chromosomes was inactivated (by imprinting).[50] This process is usually random, but they found a clear pattern, not found in straight men's mothers. This suggested that some sets of genes on the X chromosomes were being altered epigenetically – although we don't know which as yet. Simple yeast also have their equivalent of male and female forms turned on by epigenetic signals, suggesting that this is a common mechanism for most animals.[51]

In the last chapter we discussed chemical hormone disruptors like the fungicide vinclozolin. Taking this can also alter an individual's sexual attraction to others – at least in rats. Males whose great-grandparents were originally treated with vinclozolin, which subtly affects male sex hormones (androgens), were tested for their sex appeal by Mike Skinner's lab. Even after three generations of normal chemical-free upbringing, these males were less sexy to females, who consequently mated with them less often.[52] Perhaps this is the epigenetic equivalent of an unpleasant but permanent aftershave. Other studies of rats whose mothers were fed bisphenol found this slightly reduced the males' sexual performance and in contrast made the treated females more sexually active.[53]

While this is as yet unproven, it is possible that epigenetic changes related to our environment – whether from stress, cigarettes, diet or even drinking expensive mineral water from plastic bottles – could influence the sexual preferences of our kids and grandchildren. While male and female sexuality is mainly determined in the womb, it seems clear that it is very difficult to modify male sexuality or orientation after puberty, whereas females seem amenable to small changes all their lives. Let us now explore what drives individual variations in heterosexual sexual behaviour – lust and fidelity.

11

'THE FIDELITY GENE'

Pleasure, pain and the G spot

The day the world watched top golfer Tiger Woods crash his Cadillac into a fire hydrant in November 2009 at his Florida mansion, pursued by his wife swinging a golf club, was a moment as dramatic as OJ's televised car police chase. This was the most highly paid sportsman in the world, with hundreds of millions in endorsements from global companies vying for his friendly reassuring face – a smiling picture of integrity and honest hard work – to represent them.

The revelations that followed were the equivalent of the outing of the respectable Dr Jekyll and his evil twin Mr Hyde. The Tiger had become The Cheetah. His beautiful Swedish model wife had finally found out about his affairs through texts on his mobile phone which revealed all his mistresses' numbers. Not just a few, but at least 14, and as many as 40 regular women over a four- to five-year period. This was going on even when their children were being born – and he apparently wanted to carry on having sex even after receiving a phone call informing him that his father Earl had died.

His particular choice of women was also interesting: he went for large-breasted casino and strip-club hostesses, hookers and former porn stars. The perfect family man liked kinky sex S&M, threesomes, and role-playing. He paid them retainers and then large amounts of hush money in the failed cover-up. It was the fastest drop in popularity of any US non-politician. The carefully manicured and managed superficial Mr Nice Guy had been exposed. Tiger Woods claimed he had a sex addiction and checked himself into 'therapy' for 45 days in an exclusive rehab clinic in Mississippi for counselling, abstinence and prayer while he hid from the world's media and his wife's wrath.

Sex addiction or hypersexual disorder is not recognised by most sex researchers or therapists as a genuine addiction (unless it causes genuine distress), any more than is food addiction. More traditional addictions, unlike sex, have health consequences, like heroin with its problems of tolerance, overdoses and withdrawal. After Tiger and other celebrities like the actor David Duchovny checked into $40,000-a-month rehab clinics, the sex-addiction business is now booming in the US. But is it really a disorder?

Infidelity and promiscuity is perhaps the normal state for men, without the cultural constraints of marriage. This is seen in many homosexual men's promiscuous behaviour. Men of enormous wealth, power and ego like Tiger may feel untouchable and so lack any of the usual cultural controls. After all, he can still afford the estimated $500 million alimony payments. Nevertheless, although all men are programmed to some extent to seek sex and spread their genes widely, there is still a wide range of normality in sex drives and resulting promiscuity. A proportion of apparently healthy men class themselves as asexual. Around 1 per cent of the UK population surveyed in 1994 said they had never been sexually attracted to anyone – and a third of them were men.[1] Women's attitudes to fidelity and promiscuity are if anything more variable and complex.

Nicky and Louise are identical twins of 30 who are both successful professionals. Nicky is a dentist, Louise a pharmacist. Both are slim and attractive. They look very similar, although they have dyed their long dark hair differently. They have lived closely together most of their lives, except for some time at university between 18 and 23. They share many habits, like the same foods, clothes and drinks, going to the gym and jogging. Neither enjoys dangerous sports, cigarettes, drugs or gambling.

Where they differ most is in attitudes to men. Both are single, but Nicky has had five sexual partners and Louise 25. They have a similar theoretical taste in men, rating the same film stars such as Clive Owen and Johnny Depp and even Sean Connery. Louise said she was attracted to four of her sister's ex-boyfriends. Nicky in contrast said: 'I couldn't contemplate sleeping with any of my sister's many choices. I never thought they were right for her.' Louise admitted to having been unfaithful on a few occasions, and apart from once, she didn't now regret it; Nicky said she could never do that and stayed with her first boyfriend for six years, although looking back she admitted: 'I was stupid to stay so long. He didn't value me enough and wasn't right for me.' Louise admits to being impulsive in love. Nicky is very cautious, not wanting to get hurt.

Both enjoy sex and say they have orgasms, they both believe in God in some form and attended a Catholic school and were taught equally about guilt. The twins believe that one life event may have changed them or accentuated some small differences. They couldn't recall many early events, but when they were 15 they found out their father had been secretly keeping a mistress for several years. Nicky took this very badly and resented her father and the pain it caused their mother, whereas Louise, although also upset, took a less emotional view. Nicky explained: 'I could never be unfaithful or go out with a man in a relationship, as I feel bad for the other partner.' Louise has no such regrets: 'I'm choosy with my men, but I've been out with attached or married men and it's not the ideal scenario, but I realise I'm more pragmatic and not going to change the world or change men. I think my sister should have more fun – she thinks about consequences too much.'

We will never know exactly why these twins who are otherwise so similar react so differently to men in terms of fidelity and number of sexual partners. Psychologists often stress the importance of father–daughter relationships. Clearly the two girls, like other twins we have discussed, cannot have had exactly the same bond with their parents, and the balance would differ subtly between them. Nicky appears now to identify more with her mother and doesn't want to risk being hurt. Louise is more prepared to take emotional risks for pleasure. It is also possible that these personality differences were already in place when they were infants. Louise was the naughtier twin who took slightly more risks, while Nicky was more cautious. Evidence from recent studies in a range of animals and functional brain scans in humans suggests we have a personality type related to different levels of reflection before action. This is called SPS (sensory processing sensitivity), which like all personality traits has a moderate genetic component.[2]

Someone with high SPS will tend to reflect on an action much more before proceeding, while in the same situation someone with low SPS would act first and be opportunistic. These categories are akin to the plasticity genes we discussed before when talking about childhood responses to violence and abuse.[3] As these two identical twins behave differently we would predict that the key genes influencing SPS would be altered epigenetically in one, making Nicky more sensitive and emotionally distractible.

Having a thoughtful, wandering mind may be protective in some situations but not in others. The Harvard psychologists Dan Gilbert and Matthew Killinsworth used an iPhone app to record at random times of the day and night the thoughts and happiness ratings of 3,000 volunteers.[4] They found the activities that gave the greatest happiness were those where the subject was least distracted by other thoughts. Of these, love-making easily came top, although even then, 30 per cent of participants had their minds elsewhere.

According to anonymous surveys like the US General Social Survey and our own UK twin surveys, around 1 in 4 adults in Western countries admit to infidelity and probably even more practise it. Rates in men are slightly higher than in women (women may be more discreet), but are slowly rising in both sexes.[5] The fact that these traits (both infidelity and promiscuity are linked) are so common, despite the downsides of being caught – resulting in shame, divorce, injury, and sometimes death – probably has an evolutionary basis. Women who chose an infertile, violent or unsuccessful long-term partner ideally need an escape route for their genes. This is why male-dominated societies try to suppress women's natural instincts to keep their options open. We found the infidelity (or fidelity) trait has some genetic basis in humans, with around 40 per cent heritability in females[6] – similar to heritability for divorce, which can be a wanted or unwanted side-effect.

Cheating birds, lonely worms and Casanova voles

Before the age of genetics, birds were thought to be very faithful, but genetics has revealed the sad truth, and it turns out that female bird infidelity is common and heritable, being passed on from females to male offspring. Females that take a risk and raise offspring from a brief affair produce on average fitter babies due to greater immune diversity[7] – that is if they don't get caught. Males, however, spread their chances (and plentiful sperm) wider and consequently the successful ones produce greater numbers of offspring.[8] The same evolutionary drives for better genes seen in birds are likely to be operating in women, both balanced by the benefits of retaining a good male provider.

But as infidelity and divorce are in part heritable, if your parents were unfaithful or unreasonably faithful, are you destined to follow the same path? The large number of discordant twin pairs for fidelity and divorce suggest not. Culture and personality also clearly affect how satisfactory the relationship is perceived as being.[9] Animal studies can also shed some light on the mechanisms and the ability of individuals to change.

The closest that worms get to socialising, cuddling and sex is eating. A slight modification to a single gene (neuropeptide Y) can make naturally sad solitary worms start feeding sociably in groups.[10] In certain types of voles, scientists have been able to change genetically embedded behaviours. One species of vole, the montane variety, are naturally promiscuous, unfaithful, and dreadful fathers.[11] The researchers in a clever series of studies managed to alter the gene responsible (the vasopressin receptor gene) with hormones and gene therapy and bred new voles with the altered gene. They turned the naughty Casanova male voles into perfect faithful husbands and fathers.[12] These studies suggest that some of these marked social and sexual changes are actually due to very subtle changes in genes affecting the brain reward centre in the hypothalamus.

The neuropeptides vasopressin and oxytocin, the two main genes associated with bonding in voles and other mammals, are also important in humans. Oxytocin, the so-called cuddle hormone, is well known in humans to be released during childbirth and to be important in mother–child bonding. The genes as always don't work in isolation and are dependent on other hormones for their effect. Like all genes they can be up or down regulated by epigenetics.[13] In the fond prairie vole dad, the vasopressin genes are greatly over-expressed just after mating, so helping form the bond with his new girlfriend and making him hang around her and their future offspring. This mechanism of bonding is probably due to epigenetic changes in the genes from chemicals or pheromones. In humans genetic and epigenetic variations in the vasopressin receptor and oxytocin receptor gene have been linked to differences in a range of bonding behaviours – male altruism, marital status, pair bonding and autism.[14]

Although montane voles and unfaithful humans clearly differ in a few key characteristics, the general mechanisms may be similar. This would suggest that leopards can sometimes change their spots, and even the chronically and genetically unfaithful can reform, if the chemicals or potions are right. The potential to change emotions and behaviours by chemicals has not been lost on the pharmaceutical industry. A recent randomised placebo-controlled trial of 76 male volunteers found oxytocin nasal spray given 45 minutes before a task raised levels of sensitivity and empathy in males to female levels.[15] You can now buy one of these love potions – oxytocin spray – on the Internet and try it out on your male partner. Although there are some caveats. The quality of these products is unknown, and it may be hard to surreptitiously and casually administer the nasal spray to your loved one every two hours without suspicion. Interestingly, a chemical substance in common use is also able to increase oxytocin levels – and potentially empathy. It's called MDMA – more commonly known as the party drug Ecstasy. However, long-term use causing brain and liver damage is unlikely to produce the perfect empathetic father.

Let us now see how individuals vary in the sex act itself and the holy grail: orgasm.

Orgasms and restless legs

Ruth Byron was a 50-year-old housewife from Blackpool with an interesting problem. She couldn't stop having orgasms. She could have up to 100 a day. Ruth said: 'I can have ten or more orgasms at night and then have another one on the traffic island I cross as I walk the children to school. The trigger is sex with Simon, and that can set up forty or more orgasms a day. They are growing more and more intense. That is why I sought medical help.'

Simon is her ex-lodger, lover and now husband, aged only 22. He said: 'I can't deny it, at first I felt like a stud. But as the weeks go by I sometimes wish it would end. We tried a sex ban so we could get some rest, but that only lasted a night. Some might say this is the best thing ever, but it is not – it rules your life.'

It certainly messed up Ruth's life. The slightly intimidating lady's first husband left her – perhaps to get some rest after producing ten children. Then when she was discovered by the British media and billed as 'Britain's most orgasmic woman' the benefit fraud office took notice. They found out that she had been claiming her young lover as a lodger, and as well as being sentenced to community service she had to repay £6,000 in benefits. She did manage however to get some medical help for her problem, and was saved by the menopause from having further children.

Ruth's condition was only really properly recognised in 2006, when it was called persistent genital arousal disorder. Of 23 women sufferers studied recently in the Netherlands, the majority also reported other problems such as restless legs at night, and sensitive bladders.[16] Over 80 per cent couldn't wear tights or tight underwear, and even sitting provoked orgasms in most of them. The women are not hypersexual as in nymphomania, but often feel a strong desire to relieve themselves sexually. Their male partners often feel totally dominated, and have difficulty coping. When the investigators examined the women for the study they found that touching anywhere near the genital region produced an instant and embarrassing orgasm. A key diagnostic test is that the nerves supplying the genitals are over-excitable. The fact that most of these women also suffered from restless legs syndrome alerted the researchers that there must be a common mechanism that made them rename the syndrome 'restless genital syndrome'.

Restless legs syndrome (RLS) is an odd condition that unlike Ruth's is quite common, affecting up to one in ten people, with a wide range of severities. It causes an unpleasant sudden urge to move the legs and is associated with odd sensations like an itch you cannot scratch. It occurs usually at rest or at night and can be quite disturbing, needing drug treatment in some. It is about 60 per cent heritable and genes associated with it have surprisingly been easy to find.[17] These included traditional genes controlling brain activity as well as a special type of genetic material: non-coding RNA. These are small molecules that used to be thought of as 'junk'. There are tens of thousands of them which we now know are far from being rubbish. They are another important player in epigenetics and act by attracting enzymes that turn gene expression off and also by causing imprinting (turning off one copy of a gene region of a chromosome).

These two overlapping unusual Restless syndromes in the legs and genitals suggest that the sensitivity of nerves is important in the control of orgasms, and these are in part under genetic control. But orgasm is not just a question of local nerve sensitivity.

The two identical twins Jacquie and Margaret look very similar but feel they are very different. They were born 42 years ago and raised in South and East Africa with a tough Scots father and religious South African mother. 'We were raised strictly to believe in no sex before marriage and never to let someone bully you into doing something you didn't want to,' explains Jacquie. 'We both became quite religious as teenagers, although my faith lessened recently and I am no longer practising. I was always the more difficult twin with my parents, while Margaret was more accepting and easy-going – although I think she is now much more stubborn and argumentative.' The twins have never discussed sex or relationships together.

Jacquie recalls: 'My own experiences started late. I started exploring my body and masturbating around sixteen. I started seeing boys around eighteen, and had been going out with a boy for a year or so and it was going nowhere and we were about to break up, so at age twenty-two I slept with him to keep it going. It was a mistake and a big disappointment. I hardly felt anything at all.' Jacquie later married a local African and she realised in retrospect that her first lover had possessed 'a tiny little boy's penis'. Her sex life improved quickly as she got more sexually involved and lost her inhibitions. She soon found having orgasms fairly easy and could have several in a row while stimulating herself. She thought she was helped by probably having a G spot.

One day she came home to discover that her husband was having a virtual affair, via the Internet. 'I found out he was also into Internet pornography, and I just couldn't accept this. I realised instantly our marriage was a sham and was over.' Since then, although not officially divorced she has had sexual relationships and slept with three other men. The last one was a one-night stand she initiated – a friend who was staying the night. 'Unfortunately I couldn't climax, but I didn't regret the experience.'

Her sister Margaret is still very religious. Early sex was not really an option. Her husband was her one and only boyfriend. They started going out when she was 23 and waited three years until their wedding night for her first sexual experience. 'I waited partly for religious reasons and also as I wanted to discover the world a bit. In the end, it wasn't that great. I wasn't really in the mood and felt vulnerable – but happy to get it over with.

'We had sex at least four times a week for the first few years – but less now – but I never in the thirteen years of marriage had an orgasm.' Until recently, at age 42, Margaret didn't know what an orgasm was. She had tried masturbating a few times in her life, but 'nothing ever happened'. She didn't think she had a G spot. After watching a recent TV documentary about other women who had sexual problems, she tried masturbating again after a break of 20-odd years. 'It was amazing. This time something just clicked, whereas before nothing happened. I thought I had a permanent biological impairment.' She has yet to have an orgasm with her husband: she describes his technique as 'a bit mechanical' and hasn't had the courage yet to discuss her recent orgasm experience.

Why don't 1 in 6 women (like Margaret until the age of 42) ever experience orgasm, while 1 in 12 are multi-orgasmic and have them easily? Although 98 per cent of men manage to have orgasms during penetrative sexual intercourse, only a third of women do so without manual assistance. Yet this large female variation in orgasmic ability has been shown to have a clearly heritable component. In one of our most quoted twin studies (usually in women's magazines) identical twins reported more similar patterns of orgasm either by masturbation or intercourse than non-identicals and produced a heritability of 30–40 per cent.[18] An independent Australian team found virtually identical results at about the same time.[19] Could the differences between twins be related to the so-called G spot?

This is another part of our work that attracted the greatest publicity – for perhaps the wrong reasons. The G spot, beloved topic of women's magazines and sexologists, was apparently discovered by and named after Dr Ernst Gräfenberg in the 1950s, and reinvented in the Eighties, when it was linked to another mysterious phenomenon, female ejaculation.[20] In our survey 56 per cent of UK women believed that they had one – namely a discreet sensitive spot on their vaginal wall that increased arousal and sometimes orgasm. But we found no greater similarity in identical twin pairs compared with non-identical pairs, thus confirming zero genetic or family influences.[21] You will have worked out by now that virtually everything we have tested in twins has some genetic component.

Although it's very hard to prove the non-existence of something, we concluded that as the G spot is lacking in academic credibility among gynaecologists, has not been found by scans or anatomists, and had not the tiniest genetic influence, it was probably a figment of modern imagination. It is more likely an area through which the base of the clitoris can be felt and stimulated in some women. Our conclusions were not popular. We got many angry letters from Italian and French sexologists who charge patients to find their hidden G spots and from plastic surgeons who increasingly do lucrative enhancement surgery by bulking up 'the spot' with injection of fillers like collagen.[22] We also received outraged letters from 'male expert lovers' who claimed to have satisfied many women by uniquely being able to find their G spots. Strangely we didn't receive a single letter from a woman.

As the female orgasm hasn't been proved to have an important reproductive or survival role, why would there still be genes for such a clearly (though controversial) non-essential human trait? Theories include its role in reward for having more sex and therefore more babies. However, if this were the case, women with orgasms would have more babies and the genes would become more frequent, slowly taking over from women without orgasms. As we know, this hasn't happened, and women who regularly orgasm are still in the minority. This suggests that despite the dramatic effects it can have on some women (and on the sales of women's magazines) the female orgasm is just an evolutionary fossil – the counterpart of the redundant male nipple. This fossil theory proposes that the clitoris developed in the fetus as the vestigial penis before sex differentiation and subsequently got somewhat ignored thereafter in females. It was presumably easier for nature and evolution to keep the shared mechanisms of development of a 'dummy penis' in females than to devise something completely different. If so, could it in time, over generations, slowly disappear completely?

Male nipples as well as clitorises may also be on the way out. The brain area where sensory input is represented appears different for male and female nipples, with the female nipple occupying a very prominent area right in the centre of the genital area, while the hapless male nipple now receives little brain space in the boring, relatively insensitive chest region.[23] Don't get too worried. Evolution will probably allow another few million years before female orgasms disappear, and men's slowly shrinking Y chromosomes are expected to die out (taking all males with them) well before then anyway, in the next million or so years.[24]

Going back to these orgasm-susceptibility genes, how could they be acting? Certainly on aspects of personality, which we know are important. Anatomical differences could also signify. We discussed previously the effects of testosterone in the womb, mainly on the male, but when females are exposed to high levels, as in certain diseases (like congenital adrenal hyperplasia), increases occur in the size of the genitals and clitoris. In males the range of size of penises is quite limited, for good genetic reasons: too small they get laughed at, too large they don't fit and so can't transmit genes. In females the size of clitorises varies enormously. Some can be barely measurable at 5 mm and others 35 mm long, resembling a small penis. The nerve supply appears to be proportional to the size, with the smaller ones having a very dense, highly concentrated system of nerve endings.

Research on whether this matters to female sexuality or pleasure is virtually non-existent. When my former PhD student Andrea Burri designed a self-rating questionnaire and diagram with instructions on use of a mirror and tested it on our British twins, most women were quite upset just at the idea, let alone willing to do any clitoral measuring. One UK study in anaesthetised women did manage it. This showed great variability in dimensions, with three times more clitoral variability in size than for vaginal depth or penis length,[25] supporting the view that the clitoris was just an accident of evolution.[26] However, if it really is merely an accident, it is odd that the brain has reserved a special part of its sensory detection process to the clitoris in the genital area, and that unlike most males nipples it can have such dramatic effects on the brain, emotion and behaviour.[27]

Another anatomical factor influenced by hormones and development that could affect female orgasm is a variant of the anogenital distance (the clitoral–anal distance of the perineum) which is related to fetal testosterone levels. Dr A. E. Narjani, an early French sexologist, proposed in 1924 that women with a short perineal distance of less than 2.5 cm (the range is 1.5 cm to 4.5 cm if you want to measure yours) should be able to have orgasms more easily during intercourse. Data from two studies collected over 80 years ago was reanalysed statistically and showed that the shorter distance correlated with ability to have an orgasm through penetration alone. It wasn't clear if this was due to more direct contact with the clitoris externally or internally. It did however suggest that less androgen exposure *in utero* (and relatively more oestrogen) actually increased female orgasmic ability by altering anatomy.

Dr Narjani was in fact a pseudonym for Princess Marie Bonaparte, the great-grandniece of Napoleon Bonaparte, who married the Prince of Greece and Denmark. She was independently very wealthy (from Monte Carlo real estate), a student, patient and benefactor of Freud's, who helped him escape from the Nazis. She was fascinated by her own sex drive and had several high-profile affairs, among them with the French prime minister Aristide Briand. She was persistently frustrated however by her lack of ability to have orgasms through vaginal penetration alone (which her mentor Freud convinced her was the only true orgasm). She believed in two types of frigidity, one of them amenable to psychoanalysis and the other anatomical. She, along with an Austrian surgeon, devised an operation to move the clitoris closer to the vagina. They performed the operation in six women – one of them herself – but sadly to no avail. Rather than disproving her theory, failure probably stemmed from damage caused to the nerve supply during the surgery. Nevertheless, after this brave failure she focused more on the psychological aspects of female orgasm.

Andrea Burri and Lynn Cherkas from my group performed one of the few comprehensive studies of personality traits and ability to orgasm in normal women.[28] We found that when women had certain character traits such as obsessional behaviour, over-tidiness, anxiety, and a lack of so-called 'emotional intelligence', they complained more of sexual problems and difficulty achieving orgasm. When we published these results, the then best-selling *Sun* newspaper helpfully gave it the headline 'Brainy Birds Bonk Better', which, sadly, didn't precisely reflect our findings. We didn't measure IQ, but did measure the ability to empathise with others and also to register one's own feelings.[29]

Emotional intelligence is part of the empathy spectrum we discussed previously as part of the autism spectrum, and involves mirror neurones. So for women, feeling relaxed and being able to easily read their partner's (and their own) feelings and emotions are very important in achieving orgasm, whether by masturbation or intercourse. This underlines the huge gulf between men and women, and the enormous complexity of control factors underlying female sexuality.[30] An analogy in contrasting the sexes is to compare the on–off switch of a light switch, for males, with the cockpit control panel of a jumbo jet for females – minus the auto-pilot.

Wonder drugs, Viagra and the brain

Our twins Jacquie and Margaret's cases show that even with near-identical genital anatomy (as in twins), what goes on in the main sexual organ – the brain – is clearly crucial. The pharmaceutical industry has found to its cost that women are much more complicated. Whereas a simple mechanical drug acting on blood vessels like Viagra transformed many men's sex lives and made billions for the companies, for women – apart from giving them flushed cheeks – it didn't work better than placebo. Attempts with other drugs have cost the companies billions in failed trials and attracted criticism.[31]

The latest hope was flibanserin, trialled by Boehringer,

originally an anti-depressant acting on serotonin levels, which had the side effect of increasing libido and sex drive in many women – one of the commonest female complaints. It did actually work in placebo-controlled trials – it increased 'sexually satisfying episodes' (SSE) in the drug group.[32] However, the catch was that on average women improved by only just over half an SSE event per month more than with the placebo tablet. The FDA in the US deemed in 2010 that half an orgasm was too few when women (unlike men) had to take the pill every day and it had other psychological side effects in one in six women. They wanted more expensive tests and the conservative company decided to drop it. Currently a US company has bought the licence and is trying to see if they can resurrect it with some more trials and get approval with a few more SSEs per woman.

One of the only licensed drugs on the market for increasing orgasm by heightening desire is a testosterone patch giving women one extra SSE per month, but it is only used after hysterectomy and has side effects. Although most of the Big Pharma companies have given up further sex research, a few smaller ones are experimenting with nasal sprays of oxytocin, which has been used in awkward social situations, autism as well as libido and mate preference, without any conclusive results as yet.[33] These expensive failures suggest that a single magic bullet to address female sexual problems is not going to work, as so many processes are operating differently in each woman.

There is one substance that has been universally successful in all the ten or more female trials, but that so far hasn't been allowed to be commercialised by the authorities, despite it being completely safe. It increased desire and orgasm rates significantly by around 50 per cent. It is called the placebo. This emphasises the importance of the brain, emotions and expectation, and explains the continued success of aphrodisiac potions like 'Spanish fly'. As well as female emotions, some parts of the external environment are also hard to fully control – among them men. Satisfaction with the partner is a major factor, for a number of reasons. Women may sometimes have to take matters into their own hands. I spoke to a number of sex counsellors, including Julia Heiman, the director of the famous Kinsey Institute in Indiana, about how many of the 10–15 per cent of otherwise normal women who have never had an orgasm would respond to intensive therapy. She believed that with behavioural therapy and practical sessions on masturbation from an expert, the vast majority of women would eventually experience one – if they persisted long enough.[34] She believed it was more a question of threshold and sensitivity, rather than a yes/no phenomenon. Some women can orgasm just by fantasising; others may need an hour or two with soft music, vibrators and several gin and tonics.

The twin case-studies and the research around the topics of sexuality, fidelity and orgasm clearly show that the key controlling organ is not between your legs but within your head. This can be modified early on by genes and hormones influencing the brain's delicate development, and – although this is not yet proven – probably by external agents like hormone disruptor chemicals. After puberty it appears harder to change behaviours and habits, but clearly there are many examples where outside influences have modified the genes. The final story of 'anorgasmic' Margaret shows that even after 20-odd years, patterns are not hard-wired. Watching a simple TV documentary programme, together with a realisation that it is possible to change, meant that she experimented and managed to experience her first orgasms at 42 years of age.

The latest brain-scan research is beginning to show us how orgasms, emotions and brain structure may be related. Two research teams from the US and Netherlands have been putting intrepid volunteer females into functional MRI scanners and asking them to either masturbate or be manually stimulated by their partner – while keeping their head very still. While the women who are both able and willing to do this are unlikely to be very 'average', they do provide fascinating results. Both studies showed that the act of thinking about sexual stimulation, being physically stimulated and then the orgasm itself lit up over 20 common areas of the brain. Upon orgasm the blood flow was increased, but generally in the same areas. These include brain areas for touch, reward and pain. What was of great interest to the New Jersey team was the region that was activated first: the prefrontal cortex (PFC). This site in the front of the brain is involved in consciousness, self-evaluation and empathy; it had not been seen before in this context. The PFC area actually lit up more strongly when masturbation was imagined rather than carried out, suggesting that this could be the key fantasy centre and showing that those mysterious mirror neurones are involved yet again.[35]

Involvement of a small part of the PFC called the OFC (left orbito-frontal complex) was highlighted by the other Dutch group.[36] They saw significant deactivation in this area during orgasm and speculated that it needed to be switched off first for the woman to relax and have the orgasm. This would fit with the findings that personality traits like anxiety, introversion and lack of empathy make it much harder for women to orgasm. The other interesting area that lit up during orgasm was the anterior cingulate gyrus, which is important in pain control. Surprisingly, there are quite a few links between orgasm pathways and pain, involving the nerves and brain pathways.

A rare but well-reported event in some women at the crucial moment of childbirth is an unexpected orgasm – so-called 'orgasmic childbirth'.[37] Some believe that the orgasm may have a role in preparing women for childbirth and lessening the pain, but there is no real data on this. Most women however would see this as rather poor preparation, as they don't generally recall childbirth as an orgasmic experience. The true mystery of the orgasm is underlined by studies showing that women with complete spinal cord paralysis can still have real orgasms when skin near their injured area is stimulated. Or other studies showing that vaginal self-stimulation with or without orgasms can raise pain thresholds by 40–100 per cent. In rats, giving neuropeptides like oxytocin facilitated the theoretical equivalent of an orgasm as well as working as a pain reliever.[38]

Pain researchers have for a few years been using cognitive feedback measures and brain scans to help patients with chronic pain visualise how their different brain areas light up in sequence. The sex researchers hope to use brain imaging in the same way in women who can't achieve orgasm. We don't yet have any human data on the role of epigenetics and orgasms, but in a recent ongoing experiment we have taken 25 identical twin pairs with extreme and discordant pain thresholds and tested every gene's methylation status at over 20 million sites. We have found four key pain genes that are switched on epigenetically in one twin and inactivated in the other.[39] This gives great hopes for new pain therapies and suggests that the same will be true for orgasm differences.

While we are slowly uncovering the mechanisms of pain, maybe it is just as well that we will never totally understand all the mysteries of the female orgasm. Comprehending its complexity could, however, be very important for some women. What turns you or your genes on or off is clearly a key part of our sex lives.

12
BACTERIA GENES

Bugs, poo and you

In the nine months you were resting in your mother's womb you had a comfortable life free of stresses and infections. Your gut was pristine: completely sterile, with no bugs or bacteria. Your skin was covered in a protective oily coating and there were no bacteria or parasites hiding on any skin surface or in any of the nooks and crannies of your body. At birth you were suddenly released into the infested world outside. Within a few minutes even in this supposedly sterile hospital delivery room, bacteria from your mother and the midwives had begun to colonise your skin, anus, vagina and nasal passages. On the first reflex suck of your mum's nipples, bacteria started to flow into your mouth and down into your stomach and intestines. These bacteria grew and spread rapidly throughout your body. Within a few weeks your smooth baby skin was covered and your intestines swarming with millions if not trillions of different microbes. Amazingly you survived this onslaught. In fact this is a completely normal process that happens to all babies at birth and shows the intimate relationship we have all our lives with bugs. We are only just grasping how this special relationship influences our lives.

Bacteria: friend or frenemy?

Stomach cancer is usually fatal and kills about a million people a year worldwide. It has, however, declined rapidly in the

West over the last 50 years from its status as the major cause of cancer deaths in the 1980s, when it even beat lung cancer. In some parts of the globe like China and Korea, change has been slower, and stomach cancer is still a global problem.[1] Yet despite hundreds of unlikely dietary theories – ranging from what type of bracken or grass sheep grazed on before they were eaten, to changes in drinking habits – the phenomenon perplexed epidemiologists for years. The only clear associations until recently were the usual suspects, cigarettes and alcohol.

At the same time as stomach cancer has been decreasing, stomach (peptic) ulcers have steadily increased. They were until recently seen as a twentieth-century disease related to stress, causing acid release and ulcers in the lining of the stomach. Stomach cancer was first reported only in 1835, fourteen years after Napoleon Bonaparte was supposed to have died of it. This was deduced after further autopsies carried out in 2007 showed him to have multiple stomach ulcers and invasive stomach cancer – a lethal combination even without the arsenic he was being fed (according to the French) by his British captors.[2] Those painful stomach ulcers may have been the real reason his hand was always inside his coat.

Peptic ulcers and their treatments only really grabbed doctors' attention in the 1940s and 1950s, when they often perforated or bled and could be fatal. There is a long list of celebrity deaths from perforated ulcers, including James Joyce, Pope John Paul II, Ayatollah Khomeini, Alfred Nobel and Charles Darwin. The pattern of the disease has also changed in Western countries. In the last century it mainly attacked the stomach. Nowadays the sac below it, the duodenum, is a much commoner target.[3]

Once doctors had worked out that the stomach glands produced acid, an overproduction of gastric acid caused by stress or bad diet was the obvious cause of the problem. This led in the

1970s to the development of some clever drugs that reduced acid secretion (called H2 blockers). However, as soon as the drugs were stopped the ulcers returned – not so good for patients, but great for the Pharma industry, whose drug sales, even by the 1980s, were making $3 billion a year. The only other treatments were crude surgical procedures removing large bits of the stomach and cutting the vagus nerve. They were expensive and risky, with many side effects, and not usually very successful, but they kept many influential surgeons gainfully employed.

Sometimes maverick individuals who fight the prevailing medical view can be a nuisance and sometimes they make amazing breakthroughs. Barry Marshall was a doctor in Western Australia who didn't believe that stress caused ulcers, and he wanted to prove it. One day in 1982, without telling even his wife, he took a lab culture plate dish where he had grown millions of bacteria from a patient riddled with ulcers and poured the liquid into a small cup. He added some chicken broth to make the revolting brew semi-palatable and took a swig. 'I thought I might develop ulcers in a few years, but after just two days I lost my appetite, was sweating, nauseous and felt unwell. By day five, I was vomiting clear fluid which strangely (because of the bacterial infection) had lost its acidity and I also had dreadful bad breath. Eventually – after ten days – my wife, who had noticed my deteriorating health, found out, and was not happy.' She made him have an endoscopy and take antibiotics four days later. The samples taken from his stomach showed many dead bacteria, masses of defensive white cells, and damage to the stomach lining.

Barry Marshall's research with his pathology colleague Robin Warren in 1982 had shown without any doubt that *Helicobacter pylori* (HP), a curly bacterium found only in humans, was the cause of peptic ulcers. They published their findings in *The*

Lancet in 1984. 'It took us another ten years to change the fixed opinions of other doctors who treated stress, acid and ulcers like a religious belief.'

Marshall and Warren from Perth were awarded the Nobel Prize for medicine 20 years later in 2004 and treatment for chronic ulcers was changed for ever. Simple specific antibiotic treatments now cured most ulcers. They estimated globally saving 500,000 lives a year from bleeding and perforated ulcers. But as we will see there may be a price to pay for this success.

Since the branding of *Helicobacter pylori* as the primary cause of ulcers, we have seen a massive leap forward in our understanding of the world we share with microbes. We carry within us all – in addition to the 25,000 genes contained in all our cells – a large number of other mysterious passenger genes. In fact we have twenty times more non-human genes: over 500,000 different bacterial genes in our intestines and feces, coming from as many as 100 trillion microbes. Genetic technologies in the last five years mean that we can now identify these microbes by sequencing their identifying sections without having to wait weeks to grow them in culture plates. The gene sequence of each microbe is called a microbial genome, and when all are combined is called a metagenome.

In the UK it takes days or weeks and costs the NHS about £10 to grow bacteria in culture to identify them. It can now be done within minutes and for the same money with a gene sequencing machine, and will revolutionise how we detect infections routinely. It is already having a major effect on investigating the origin of disease outbreaks. Examples include the proof that the recent German salad *E. coli* outbreak didn't come from mutated bacteria on contaminated Spanish cucumbers, and that 'careless' surgeons were not the true sources of superbug infections such as MRSA (Methicillin-resistant *Staphylococcus aureus*) or *Clostridium difficile* in the UK.[4] Epidemics from diseases like TB or influenza can now often be traced to the exact date of arrival of a single human 'super infector' by sequencing the exact genetic mutation of the bacteria or virus.

Helicobacter pylori, the recently discovered bacterium that causes ulcers, has co-evolved with humans who are its only known host. It has a pretty depressing life, has never seen the light of day and can't live outside the human gut. About half of 50-year-old Americans now have the bacteria in their gut – but these seem only to cause ulcers in about one in five people.[5] Rates vary around ten-fold across the world, and Korea reportedly has one of the highest rates of over 90 per cent. The HP bacterium has changed and adapted along with humans, and we believe it travelled with us out of East Africa around 80,000 years ago and is actually a 'friendly bacterium' in most cases. As we are its only host it has no interest in harming us. Surveys of stored historical fecal and blood samples suggest that the levels we carry have been dropping, even before we started trying to eradicate them to prevent ulcers. Rates in the US based on stored blood show a 50 per cent decline in HP since 1968. This is likely due to better hygiene and less transfer and mixing of species of bacteria from person to person – which often happens in early life.

Hygiene, allergies and your gut

While hygiene may have reduced some infections, it may have a downside, as suggested first by David Strachan, a clever epidemiologist I did my Master's degree with in London. He first proposed a 'hygiene' hypothesis to explain why rural kids got less asthma and allergies.[6] He suggested that overzealous cleanliness was partly responsible for the recent allergy epidemic since the war. Many common infections have in the same period decreased dramatically in the West, among them measles, mumps, TB, scarlet fever, and many bacterial infections. As gastric cancer has decreased, the disappearance or recent reduction in the levels of HP in our bowels has been linked to unexplained increases in a number of other cancers. These include cancer of the oesophagus (the gullet) and other parts of the stomach (called the gastric cardia) that used to be rare, as well as a dramatic increase in allergies and diabetes.[7]

The lesson from HP is that we interfere with our gut flora at our peril. Remember HP is one of hundreds of thousands of known and unknown bacteria species that we coexist with and whose genes produce thousands of different proteins. There is some evidence that 'gut-friendly' yogurts with acid-resistant lactic acid (*Lactobacillus* and *Bifidobacillus*) bacteria given after meals can suppress HP activity in the stomach.[8] While if taken regularly this may be useful for eradicating HP-related ulcers, it may covertly be actually increasing other diseases. A recent study of 70,000 Danish pregnancies found that mothers who ate low-fat yogurts during pregnancy produced more allergic and asthmatic kids than those that didn't.[9]

There is only one reliable method of curing type II diabetes in an obese patient within a few days. Although there are many drugs that help a bit, the treatment that works fast and keeps patients cured for over ten years is called gastric bypass (bariatric) surgery. This surgery (and its milder form gastric banding) has been around since the 1950s but has only recently begun to be performed on a large scale. In my hospital in London around 500 operations are now performed each year. The media in many countries – the UK included – have usually been opposed to it because of its costs and the perception that it is used in unworthy patients who are not ill, but just lack willpower.

But it works. Within 48 hours of the operation to effectively bypass most of the stomach, sugar and insulin levels of patients return to normal and stay that way for ten years in over 70 per cent of cases.[10] The conventional explanation is that this is due to malabsorption and that nutrients just pass through the gut quicker, but there is no hard evidence for this and many patients actually have constipation, suggesting that passage of food through the gut is if anything slower. Some of these rapid changes could be due to changes in the gut hormones that signal the brain when it's hungry and full, which we discussed in chapter 7.[11] However, patients' eating habits don't alter dramatically, although they have more frequent small meals and less fatty foods. What is also happening is that the gut bacteria change rapidly and suddenly take over new areas[12] and alter metabolism.[13] This could be the reason the operation works so well. We just know so little yet about which bacteria are good or bad for us.

But there may be other ways of changing bacteria and their genes without surgery or eradicating them via antibiotics. We now know that epigenetic mechanisms are not confined to humans and also work in other organisms.

Bacterial genes, like those of their human counterparts, can be methylated and so get activated and deactivated by external factors.[14] Although they also have high gene-mutation rates, epigenetic change is another important factor in how they can adapt to new environments and possibly how they cope with antibiotics. Methylation also plays a role in the way that friendly or aggressive (virulent) bacteria behave towards the host human.[15] A recent study looked at a mouse model of stomach cancer and found that giving folic acid supplements to these mice at birth, when they have no gut bacteria, protected them against later cancer – via methylating the new bacteria in their intestines.

This process deactivated the bacterial genes and proteins that would normally kick-start the cancer process by disrupting the DNA in the cells of the stomach wall.[16]

The average child in the West takes 15 courses of antibiotics before the age of 18 – most of them unnecessary. This means that our gut bacteria, including their genes and proteins, are very different to those of our parents and grandparents 50 years ago. The drugs cause mutations in our bacteria to make them more resistant to antibiotics, but they also alter their normal protective functions. Bacteria like HP seem to act as our home defence system, actually reducing levels of harmful bacterial species. Some of this effect is through epigenetic mechanisms that influence the host's immune genes. Kids are born usually without bacteria and most, like HP, are rapidly accumulated in the first year of life.[17]

Some experts predict that we will soon be actually giving newborn babies HP infections to redress the defensive balance and reduce allergies. The change in our bacteria could also be one reason that not just allergies, but diabetes and possibly even heart disease have also increased since the introduction of antibiotics. A recent study showed that feeding young mice different amounts of the key constituents of fat altered their gut flora, which – depending on the chemicals – in turn altered the amounts of fat produced by the liver which lead to increased or decreased fatty plaques lining the arteries (atherosclerosis). This suggests that drugs that alter our gut flora could reduce risk of heart disease.[18] One of these fat metabolites was choline, which – as we discussed earlier – can alter methylation of key genes.

Other studies in humans have found that certain bacteria normally found in the gut are actually present in the damaged vessel wall of patients with heart disease – and may be partly responsible.[19] Some other diseases like dental caries (tooth decay) and periodontitis (gum inflammation) are also closely related to the bacteria in our mouths, which are also affected by taking courses of antibiotics. Although caries has decreased in recent years due to fluoride, antibiotics could explain recent increases in gum disease. These gum infections are also weakly related to increases in heart disease, although we don't yet fully understand the mechanisms.[20]

The fact that bacteria are so clever in adapting their genes via mutations or epigenetics to protect themselves against new antibiotics has come as a shock to doctors. However, a recent study has shown the bacteria have been practising survival techniques for millennia. Scientists digging in a permafrost site in the Yukon found some ancient uncontaminated DNA from bacteria that was over 30,000 years old, when mammoths roamed the earth.[21] They found these ancient bacterial genes already contained a battery of antibiotic resistance genes – which they could quickly swap with each other. They were used to signal to each other and to defend themselves against other bacteria and fungi which also produce natural antibiotics. So bacteria have had millions of years of evolution arming themselves to deal with any antibiotic we can throw at them. It seems we can kill some of them, but often only at the expense of harming ourselves. Our personal bugs have more experience and complexity than we have given them credit for.

Psychosomatic bacteria

Ellen and Eva are British twins married and in their fifties and both on the heavy side, weighing around 12 stone. Both twins were small at birth, and Ellen, the lighter, weighed only 1,400 grams (3 lb). Ellen became a vegetarian age 25 for 15 years and felt healthier and lighter, but got fed up with red peppers and tofu and reverted to white meat. Both sisters have suffered for the last ten years from an intermittent bloating feeling in the abdomen and cramp-like pain that was sometimes relieved by going to the toilet.

When their pain is bad they vary widely in symptoms – from being constipated to having runny stools. 'We both tried changing our diets, but nothing seemed to help much,' says Eva. Ellen remembers 'having less symptoms when I was vegetarian'. Both went to see their different GPs, who did a few tests to rule out infections and bleeding, and they got similar advice. 'My GP said it was probably just stress and related to depression. I was told to relax and forget about it and come back for anti-depressant tablets if it didn't improve,' said Ellen. Eva said: 'My GP told me much the same. Neither of us believed stress was the main cause and we didn't go back, as we didn't want to take those tablets.'

Irritable bowel syndrome (IBS) is a relatively modern diagnosis and is made mainly by excluding other more serious illnesses. IBS doesn't kill you but is a source of pain and discomfort for many people. Some estimates suggest it affects up to 20 per cent of the population at some time of their lives. This high prevalence and lack of obvious evidence of pathology have made many doctors believe it is a modern invention, dreamed up by greedy drug companies to try and 'medicalise' a normal range of human variation. It has been linked to other similar disorders with similar obscure causes like fibromyalgia and chronic fatigue syndrome. All three conditions have been called psychosomatic, as patients often have associated anxiety and depression, thereby creating a vicious circle of symptoms. Many patients and their patient groups sadly feel very threatened by the perceived stigma of a psychological component, and some of my colleagues researching chronic fatigue syndrome have faced death threats and subsequently changed research areas, much to the detriment of science.

Two large twin studies have been performed on IBS,[22] one by our group in collaboration with a gastroenterologist. Nigel Trudghill found that out of 5,000 female twins tested, 17 per cent had mild symptoms of IBS, but it was not clearly heritable. A larger Norwegian study of 12,000 twins used stricter criteria and consequently identified more severe disease. It found a clear genetic heritability of 48 per cent in these more badly affected women.[23] The study also found that rates were much higher in the smaller twins at birth – below 1,500 grams – suggesting that early life events were also important. The view of most doctors is still that this is not a 'real disease' and just a manifestation of anxiety and depression common in middle-aged women. But could something else be going on beneath the surface? Again discordant twins might give us a clue.

Rosemary was chatting to her friends at a birthday party lunch in the Barley Mow restaurant outside Guildford, enjoying her steak and onions with a side salad. She had drunk half a glass of rosé wine and was having fun. All at once she started getting stomach pains and felt hot and sweaty. She got up to go to the bathroom, but before she could start walking she was gripped by a vice-like stomach cramp and had uncontrollable explosive diarrhoea – just before she fainted. 'I will never forget that meal – it was so embarrassing,' she remembers, 'My friends helped me get to the local Frimley Park Hospital, where I was given a drip and antibiotics and stayed for several days.'

The doctors did all kinds of tests with endoscopy tubes at both ends in her stomach and colon, but couldn't find anything wrong. 'They asked me all kinds of odd questions about whether I was depressed and was I still sleeping with my husband. I think they were checking whether I was a bit mad.' Her symptoms settled down slightly, but her bowels were still alternating between constipation for several days with episodes of bloating, cramps and diarrhoea. She took advice from the hospital on changing her eating habits – they advised her to eat little and often – and through trial and error she cut out various foods that aggravated it. 'I found that if I ate fruit with skin, porridge, muesli, bran, nuts or granary bread, I would be running to the loo half an hour later.' Twenty years later she feels she has it under control.

Her sister Jennifer, now aged 65, who lives about 50 miles away in southern England, never had any similar problems, although they both think they remember 'sitting as toddlers on our potties for long periods of time, waiting for something to happen. Our mother said we had "lazy bowels".' Jennifer never had the constipation that Rosemary had in her life, although 'for half our life we had pretty much the same diet. Strangely, as long as I can remember I've always had a bigger appetite than Rosemary, and actually am less active. But I must have a better metabolism. Rosemary thinks I eat double what she does, but I think it's closer to 50 per cent more. She is always struggling to finish her plate and I have no problem eating any kind of food.' Nowadays, although she has been heavier in the past, Jennifer weighs around 8 stone – about a stone lighter than Rosemary. She says she also has different sleeping habits and is unable to have naps in the daytime like her sister.

What could be going on in Rosemary that is not in Jennifer, her identical twin, to cause these major differences in their bowel habits and perhaps their appetite and metabolism? What also might be similar in Ellen and Eva, to give them such similar symptoms? It might have something to do with the half-million or so bacterial genes in our intestines and the trillions of microbes. Although about 40 per cent of our own microbes are shared with most other humans, many are specific to just a few of us, leading to a potential pool of over 3 million unique bacterial genes – 150 times more variety than the human genome.[24] These genes make proteins that have a key role in many processes from breaking down carbohydrates and sugars to creating fats and vitamins. Thus we possess a huge accessory genome that can influence how we filter and metabolise our diet and derive energy from it – as well as many other processes that we can only guess at.[25]

Our gut is only one part of where bacteria and their genes hang out. Skin, urine, vagina, hair and armpits are other sites where different bacteria cluster together, the exact blend of bugs reflecting a mixture of the influence of the host human and a preference for the individual site.[26] In the 1970s NASA scientists were very interested in what happened to bacteria sent into space with astronauts, and importantly – for obvious practical reasons – how they could control the fecal waste of astronauts in the weightless environment. They found that while they could modify the bugs slightly with diet, they noticed enormous differences between astronauts which they couldn't easily alter.[27]

We worked with Cornell University to get a better estimate of the heritability of gut flora of twins. Our helpful volunteers sent us a bit of their poo on a plastic spoon and Ruth Ley ran a genetic sequence analysis of the contents to identify the range and proportion of common bacteria in each individual. So far the results on a few hundred twins show a wide variation between people, but identical twins definitely share more types of bugs than do the non-identicals. We are extending this study to next fully sequence in depth all the microbe species (the metagenome) contained in each of the intestines of our 7,000 twins. This huge task is being done with the considerable help of the largest gene-sequencing facility in the world: the

BGI (originally the Beijing Genomics Institute) in Shenzhen in China. They have invested billions, and their visionary director Wang Jun clearly sees a huge future potential in epigenetics and the metagenome project to identify and understand all the bacteria that live within us.[28]

These twin results show that bacteria can recognise and prefer the genes of certain humans to others. Some of these bacterial species prefer younger or older hosts, and as there are major differences in gut flora between young and old people, some of these bacteria may turn out to be important in the ageing process itself. Deep sequencing of 22 gut microbiomes from individuals in four countries has shown that our gut contents belong to three main distinctive groups, called 'enterotypes' – a bit like blood groups.[29] Strangely, these are not related to race or place of birth, and the groups may well behave very differently in affecting how we interact with our environment, diet, and even ageing. Each of these has some dominant bacterial species. Some bacterial species also prefer to live in hosts that produce different amounts of body fat. This may not be just chance or a passive process.

Lazy bacteria, obesity and diseases

The biggest surprise in this field came in 2006 when a team from St Louis compared the feces of a genetically fat and a genetically lean strain of mice.[30] They found they had very different bacterial species. What was more exciting is that the bacteria of the fat mice could metabolise and extract much more energy and calories from food than those of the skinny ones. They then confirmed the same result in fat and skinny humans. This means that some people carry for most of their lives some bugs that greedily grab all the calories they can from their guts

– which means that for the same amount of food the host will store more fat and put on more weight.[31] Having these microbes in our guts was probably a good thing in our early evolution, and by storing fat helped survival in hard times – but not great nowadays.

So within us all lies a parallel universe of warring personalised bacterial colonies, each of them with genes and proteins with a possible role in our health of which we remain largely ignorant. We are only just scratching the surface of this hidden universe, but so far it offers us a glimpse of their potentially huge importance. The type of bacteria and the extra genes we have in our guts could explain around 50 per cent of obesity, compared with the paltry 1–2 per cent explained by the common human genes we have discovered so far.[32]

Other common diseases could also be affected by our bacterial guests and their genes. Colon (bowel) cancer is very common and has only a 15 per cent genetic component. It varies a lot between countries and geographical regions, and red meat diet and high fat intakes have been implicated in its cause. However, studies of the intestinal flora of populations at high risk of colon cancer have shown very high numbers of *Bifidobacillus* species and low levels of *Lactobacillus*.[33] There is emerging evidence that these bugs and their genes interact closely with our own genes and proteins. They have been found to alter the leakiness of the gut wall and alter the regulation of the immune system, and even cause DNA damage in our own genes.[34] These processes are seen as crucial in many diseases, like the bowel diseases (Crohn's disease, coeliac disease, IBS), as well as other allergies and cancer. The use of tests of our gut flora for diagnostics of early disease or high risk is suddenly a big research area and could explain much of bowel-cancer risk.[35]

Diseases like IBS, as we saw with our twins, are often associated with other problems like depression or anxiety, and this is often thought to be the cause of the symptoms. However, several rodent studies have shown that manipulating the gut flora by adding or removing certain species can actually alter their mood and behaviour.[36] Recent research has shown that in mice, changing their gut flora by being given *Lactobacillus* yogurt can quite rapidly alter their mood and drive.[37] They found differences in the expression of GABA neurotransmitter receptors in the brain of the happier yogurt-fed mice. We don't yet really understand how these effects occur, although epigenetic modifications and the nerve supply between the gut and the brain (vagus nerve) both appear to be important. We are bound to find many more unexpected links between our bacteria and our immune system and brains soon.[38]

Super poo

In terms of novel treatments for diseases like IBS, you probably can't get more thought-provoking than one early-stage trial: fecal transplant, or less technically 'poo transfusion'. This distasteful procedure takes healthy gut feces from a generous donor, puts it in a blender, and via a tube through nose or bottom (you don't have to drink it), passes it on to the patient. The new bacteria make themselves right at home in the new gut, taking over from the old residents within a few days, and produce a dramatic and long-lasting reduction in symptoms.

Tom Borody in Sydney has been pioneering fecal transplants since the mid-1980s.[39] They have been shown to be an excellent cure for persistent bowel infections with the superbug *Clostridium difficile*, which kills 300 people a day in the US, that can actually be caused by broad-spectrum antibiotics.[40] It also shows promising early results in Crohn's disease and colitis and more recently IBS. It is not clear if Professor Borody has yet copied his Nobel-winning compatriot Barry Marshall in self-treatment.

The future may bring a new form of super-poo that we will be taking routinely to boost our gut defences. For those too squeamish, a new form of broad-spectrum antibiotics have been tested in IBS patients that are able to get past the acid in the stomach and target the gut flora in the colon. The trial of rifaximin was pretty successful, with 40 per cent of patients improving after two weeks' treatment compared with 25–30 per cent with placebo.[41] However, as these broad attacks on the gut flora could harm some of the good guys, the natural solution may in theory be preferable.

So if some bugs can make us lose or gain weight, give or prevent allergies or prevent ulcers and cancer, we should know how we can safely change our gut bacteria. As usual, Grandma's advice may be best. Eat more greens. This may be good for us epigenetically with the help of our bacteria. When our gut bacteria break down and metabolise these plants they produce the natural chemical butyrate, which is protective against infections and acts as a weak chemical (an anti-histone deacetylator) that could change our genes epigenetically.[42]

What is clear is that we should think much more carefully when using antibiotics without good reason, and need to be cautious with so-called 'gut-friendly' yogurts[43] or bacteria 'health' teas like kombucha,[44] which at certain times of life could be helpful, and at others just the opposite. Animal studies have shown that manipulating the diets even of larger mammals like cats can change gut flora over a few months.[45] Evidence is sparse in humans, but although differences in bacteria can be seen due to breast or bottle feeding,[46] in adults effective dietary changes probably need to be over a longer time frame.

In the future we may need to personalise our and our children's diets depending on our enterotype, or even exchange bowel contents with skinny healthy humans to modify our own bacterial lodgers. We are actually super-complex organisms made up of many different species that may be controlling our bodies and minds much more than we think.

13

IDENTICAL GENES

Clones, identity and the future

She was 'a miracle baby' conceived after a long wait and several failed attempts. People said it could never be done. Her adoptive parents were becoming frustrated. Cells had been taken from breast tissue donated by a healthy relative. These cells were in a resting state and DNA from one of the cells was extracted and put into a scooped-out egg cell without a nucleus (i.e. with no DNA). The DNA and the new egg lining were fused together by a short burst of electricity to form a single cell that started to divide within a few days. This group of cells was then transplanted into a surrogate mother. She was happy to nourish the adopted fetus until she had a normal delivery several months later and then handed her back. To general joy, the baby was born, and appeared to be normal and healthy in every way.

When she was older she managed to produce several normal children naturally. However, when she was in her early forties she started to develop pain and problems when running. This spread to pain even just walking, and she was diagnosed as having premature osteoarthritis. She deteriorated rapidly and also developed progressive lung problems and trouble breathing, despite never smoking. Lab tests showed she had shortened telomeres on her chromosomes (the markers of biological age we have mentioned before), suggesting that she was ageing twice as fast as her contemporaries. Sadly she died soon afterwards, and although she personally never spoke to the press, being such a celebrity she was deeply mourned around the world.

This birth was no ordinary IVF treatment. Rather than taking the egg and sperm from two individuals, the baby was created from a single genome of one individual – and the baby was Dolly the sheep. She was named by her creators Ian Wilmut and Keith Campbell after Dolly Parton, in tribute to the mammary source of her DNA. She died prematurely in 2003 aged only half the normal lifespan – the first successfully cloned mammal. This was the first time a differentiated adult cell (somatic cell) had been reprogrammed backwards to become a simple universal cell that could develop into every cell the body needed. It was a great breakthrough that earned Wilmut and Campbell a Nobel Prize. As so often in science, someone had in fact got there first, but without the same worldwide recognition. John Gurdon, a modest Cambridge scientist, had done much the same thing 40 years before when he cloned a tadpole from a mature frog cell and showed that reprogramming cells back to their original state was possible.[1]

Despite the excitement of the scientific world, Dolly's early death raised a number of questions about reproductive cloning. The first was how difficult it was. Dolly was the only success in 277 attempts and was incredibly costly, making this a poor method for a commonplace procedure. The second problem was the small but consistent ageing defects in Dolly that were also picked up in other animals that have been cloned since. The DNA of Dolly and her genetic mother were identical, so any differences had to be epigenetic. Something about the process of artificially creating the egg may have bypassed some of the normal processes of clearing the epigenetic signals that occur when a sperm and egg fuse.

Since 2004 attempts to clone pets by companies like Genetic

Saving and Clone have produced a few cats. Little Nicky was the first created for a bereaved Texan lady for $50,000, and BioArts International Inc successfully cloned Missy the dog. But most commercial programmes have now stopped, because of the high failure rate, early deaths and complaints by customers that often their beloved animals don't look or act exactly the same. As an example, the very first cat to be cloned was called 'cc' (for carbon copy) in 2002, after 87 failed attempts. She was a cute brown and white tabby cat, but unfortunately looked nothing like the genetic original, which was a tri-coloured calico (tortoiseshell). The reason for this was that the colour banding on the fur hairs is caused by imprinting of one copy of the X chromosome, which is naturally random. This means that because of epigenetic differences even clones – despite sharing the same DNA sequence – are not going to have the same colouring. She apparently also had quite a different personality to Rainbow the original, and it's highly likely that there were myriad other differences under the surface.

A Korean team (RNL Bio) who have cloned several wolves and dogs have been more successful, and after cloning five copies of an original pet 'Booger' for a grateful US customer are now offering the service for a mere $150,000 a puppy. In China they recently cloned five piglets from a 150 kg castrated hog (named 'Strong-willed pig') that became a porcine national hero after surviving an earthquake, buried in rubble for 36 days. But if you're not that rich and you are interested in cloning your pet – you can't just dig up Fido from the garden – you need to be well prepared in advance and use fresh or cryo-preserved cells. The good news is that for only $6,000 you can have your favourite cat or small dog cryo-preserved. Goldfish are much cheaper.

The dawn of reproductive cloning in humans was announced to the world in 2005 by the South Korean scientist Woo-suk

Hwang. After successfully working on dogs, he published a study showing that he had created embryonic stem cells from skin cells that could grow into a potential human fetus. He became a national hero. Sadly he got into trouble, first for taking eggs from the ovaries of his female research staff, and then accused of fraudulent research. He (like others before him) initially claimed his work had been sabotaged, but he was exposed and the scientific papers retracted.

About the same time as Dr Hwang was being exposed as a fraud, a Japanese group led by a young Shinya Yamanaka had made a real breakthrough after years of genuine research. He managed to shock adult skin cells back into their primitive multipurpose (pluripotent) state as stem cells using a cocktail of four natural chemical factors.[2] In other words, he turned back the clock on fully developed cells so that they 'regressed back' into an earlier state. Although these are not quite the same as embryonic stem cells and can't create a whole human being, they are amazingly useful in medicine and can in theory produce and so replace most other cells in the body, such as neuronal, pancreas and heart cells.[3] The key chemical factors used to shock the adult cells to regress are under tight epigenetic control, showing how the two processes are closely linked.[4] We have started looking at stem cells from identical twins and have already found differences in how they grow and respond.

But what about the common procedure of producing test-tube babies, also known as in vitro fertilisation (meaning combining the sperm and egg outside the body in a glass or plastic tube)? Could epigenetics be a complicating factor in that process which occurs outside the normal process of the cell? The first IVF baby was created by Nobel Prize winner Robert Edwards in 1978, and IVF is now commonplace, with millions born every year. However, there are downsides that may emerge with time. Studies show that rates of certain birth defects seen in the first year (like cleft lip, heart defects and club foot) are approximately double those expected in normal pregnancies.[5] A certain number of rare diseases, which include Beckwith–Wiedemann syndrome, are up to six times more common – diseases caused by imprinting (the turning off epigenetically of one copy of a gene or large chromosomal section).

So although most children born to IVF are healthy, clearly some epigenetic problems do occur which could cause disease problems later in life.[6] The issue is complicated by the fact that 20 per cent of infertile males have imprinting defects in certain key genes. These same sub-fertile males are likely to use IVF anyway to have kids, and so may pass on these epigenetic defects directly to their children. This is in addition to any effect of the artificial process on the genes.[7] No IVF kids have yet lived for more than thirty years – the time for which the treatment has been available – so we don't know yet whether they will in the future develop any age-related diseases associated with abnormal imprinting or epigenetics. It is however clearly a worrying possibility.

The other group that has flourished and grown thanks to IVF is twins. It is estimated in Western countries that over 25 per cent of newborn twins are due to IVF – many of them it seems these days to Hollywood celebs. The usual reason is implanting several fertilised eggs as a back-up in case one fails, so they are usually non-identical twins. Some studies have shown that IVF twins tend to be born slightly smaller and earlier than natural twins and have more medical problems soon after birth.[8] A recent Chinese study in 59 IVF twin pairs has shown greater variability of epigenetic signals than expected.[9] We are currently exploring rare identical twins born by IVF with a Melbourne team to see if they have more early epigenetic differences as they develop.[10] As we discussed previously, chemicals in plastics could interfere subtly with our genes. Some studies suggest that multipotent embryonic stem cells stored in plastic plates have more epigenetic changes than natural ones. This could explain why eggs or sperm stored in plastic could have minor epigenetic memories that stay with them during their development.

While IVF babies have been with us for 30 years, and problems are mainly theoretical, producing human clones safely is still a long way off, both scientifically and legally. But what would it be like to be a human clone? Much of our perception comes from our cinema screens. Science fiction films such as *The Island*, *Multiplicity*, *The Surrogates*, *Sleeper* and *The Stepford Wives* generally depict either doppelgangers with a shared mind and identity or robotic automatons. I asked some 13- and 14-year-olds about human cloning – they had just heard Professor Wilmut at the Royal Institute Christmas lecture talking about Dolly the sheep. They were unanimously opposed to the idea.[11] The irony was that I was asking identical twins who were all clones themselves. Other more rigorous population surveys of students show similar patterns, with many wanting to ban the creation of a theoretical belated twin born years later.[12] My view is that our sense of identity is so strong, we have such a powerful sense of 'I', that you would probably never 'feel like a clone'. Identical twins are probably more similar to each other than any artificial clone could ever be, sharing the same time and space with each other, yet they feel completely different.

When with my colleague Barbara Prainsack, who works in the developing field of the politics of bioscience, we asked the views of several thousand identical twins by questionnaire, 45 per cent said that reproductive cloning should never be allowed for any reason, even for medical purposes.[13] When Barbara probed more deeply in one-to-one interviews,[14] she found that what most upset the twins was the idea that whoever was overseeing or planning the cloning – 'the cloner' – had an ulterior motive, which would upset the clone. Whereas the identical twins were created by 'chance', and so had no 'cloner' to blame. Another fear they had was that because of their physical similarity the clone pair would be mistaken for having the same personality and therefore a lack of individuality. This mirrors a common complaint of twins: they hate being referred to collectively as 'the twins', as often happens in families and social groups. So can the experiences of twins inform us all about our own feelings of self?

One of the main arguments against human reproductive cloning is the 'loss of dignity' compromising the individuality of the clonee. Ian Wilmut, the creator of Dolly the sheep, doubts 'that a clone would necessarily have the same opportunity for individual development as a child produced by sexual reproduction'.[15] National governments were asked by the UN to ban human cloning because it was seen as 'incompatible with human dignity'. Part of the rationale underlying this claim is the concern that a person who has purposely been created genetically identical with another person is likely to suffer from greater difficulties in developing an individual identity. The UN clearly hadn't sought twins' views on the nature of individuality.

Twins on twins

Jessica and June were identical twins whose views were quite typical. Jessica said: 'People still now always look at us as one person. They don't see us as individual people.' Said June: 'People always thought if there's something wrong with one of us, there's also something wrong with the other. They think that because we're identical, everything's exactly the same. They never look for any differences. Earlier in our lives we were referred to as "the Twins" or "the Twinnies" and hardly ever as individuals. That was always very annoying.'

When we asked non-identical twins about whether they would have liked to be identical, the universal response was No. As fraternal twin Sara explained: 'Oh God, yeah. Because I believe in your own identity, you know? You've got this other person who looks like you. I find this a bit too sickly, you know . . . They're dressed the same, they've the same mates. No! It doesn't seem as if they have their own identity! There's something very unnatural about identical twins.' Yet, like all the identical twins we spoke to, Jessica and Jane would never swap places with non-identicals.

Alexis and Saffi, having grown up thinking that they were not identical twins (the paediatrician had drawn that conclusion from the existence of two placentas at their birth), found out that they were MZ twins when we tested them. Saffi describes the effect this had on her and her sister: 'To actually be told that we were actually one egg that split rather than two separate eggs made the bond much stronger, as if we had that bond anyway. Finding out [that we were identical] was the missing piece of the jigsaw. Knowing we are clones doesn't change the way we feel – we are still different people that share many things in common.' According to these MZ twins, being identical twins means being made of a different 'stuff' than just genes. In general they didn't believe their behaviour was predetermined.[16]

The fact that people don't usually wish to swap places is perhaps part of the general human phenomenon that you value more something you already have, as opposed to what you might have – the so-called endowment effect. While the vast majority of twins are very happy being twins and dismiss genetic determinism, a few twins are not so happy.[17] We have several identical twin pairs who attend our research sessions on the understanding that they never meet their co-twin, with whom they are in permanent dispute. Sometimes twin or normal sibling rivalry can be unbearable.

The Han twins Gina and Sunny were born in South Korea, but moved to San Diego as teenagers. They both did well at school, but were very competitive and from when anyone could remember were always fighting. Sunny was the first born by a few minutes and in Korean culture was favoured. Schoolmates and teachers recalled that 'although both were bright and attractive, Sunny was the more outgoing and popular'. After school they first did waitressing but then separated. Gina became a croupier, which was well paid, but she couldn't keep up with the gambling habit she had picked up. Sunny was more successful. She enrolled as a student in a university outside LA, wore designer clothes and drove a new BMW. This was all done without any obvious source of income. Some of Sunny's suspicious wealth came from stolen credit cards, which she was eventually arrested for, but managed to avoid jail. Meanwhile Gina's debts increased and one day she stole Sunny's cards and car. A fight ensued and Sunny broke Gina's nose and pressed charges, so that Gina would have to serve jail time.

Gina, driven by envy and revenge, was not a happy twin, and a few months later she was looking for help in murdering her sister. 'I want that bitch killed,' she said. She used her looks and powers of persuasion to find two gullible teenage boys to help her. They bought rope, duct-tape and some guns. While she waited in the car, the boys broke into the house posing as salesmen and tied up Sunny's flatmate Kim and then Sunny herself, and put them both in the bath with a gun to their heads. Luckily Sunny had rung 911 on her mobile seconds before the boys discovered her, and the police were just round the corner and arrested them both. The next day they tracked down her sister. At the trial Gina contended that she only ever meant to scare Sunny: 'I loved her, as we are the same blood.' The jury didn't believe her, and she is serving 26 years in jail. Her sister Sunny seemed full of remorse – 'When you think about whose fault it is for Gina's conviction, it's probably my fault,' she told the local paper. However, her guilt had worn off when she happily sold her exclusive story to the media for another $100,000. Gina is due to be released on parole in a few years and it would be interesting to sit in on the family reunion.

Though few twin disagreements are this extreme, the story of Sunny and Gina shows the possible dangers for some personalities of competing continuously with someone others regard as so similar. This rivalry and urge to be different is frequently seen also in non-identical twins and normal siblings who share only half their genes. The fact that identical clones (twins) also often feel they are very different to their clone, and that genes are not that important, perhaps gives us clues to our own views of self and our individuality. To return to the conjoined identical twins in our preface who shared a part of their bodies, they still felt very different and wanted independence even at the risk of death. It would appear we are programmed to think and be that way – so that we can be distinguished more clearly. This is why, as we saw earlier, twins tend to react more differently than expected to the same stimuli, such as parents or schoolteachers.

We have seen lots of examples where nature sets up a part of our non-crucial development to be random and variable, and this variability or plasticity can be counter-intuitively considered hard-wired. It is as if nature and evolution didn't want us to be too similar to each other – perhaps this is part of the success of the human race – and plays a role in our ability to adapt and survive when thousands of other species have failed.

Identically different

In the 1978 SF thriller *The Boys from Brazil*, which starred Gregory Peck, Hitler's ear is used to clone new versions of himself 25 years after his death. Now that we understand the crucial role of epigenetics, it looks far more likely that if Hitler's theoretical clone were alive today, and without knowledge of his progenitor, he would be a grumpy retired painter and decorator, not an evil tyrant. Without the trademark bad haircut and small moustache it is unlikely that he would even be recognised, although the facial features and mannerisms would probably be close.

The title of this book is *Identically Different*. We have seen how genes closely determine our anatomy and looks, and to a large extent most of our unconscious mannerisms, such as the way we laugh or drink a cup of tea. Humans are very sensitive to subtle facial clues and body language to help avoid danger and pick sexual partners. In this way strangers think twins are spookily alike, and often brothers and sisters will appear to a stranger more alike than they or the family perceive. But superficial appearances and mannerisms cloud our views and exaggerate the very real differences in identity and personality. The genes that ensure our flexibility and randomness (plasticity genes) may be protecting us in subtle ways. They ensure greater variation and unpredictability in the characteristics of each generation when exposed to changing environments. This variation is crucial to human survival: without it we might all perish by reacting in the same way to famines, epidemics, gluttony and natural disasters.

When we are young and our quadrillion or so nerve connections are getting established, some neuroscientists believe that each brain network starts to behave semi-autonomously, rather as if each group of brain cells was evolving like a natural animal species – so-called Neural Darwinism.[18] This would help explain how even identical brains develop differently as groups and networks of neurones expand by a trial-and-error method. Remember also that memories may actually be formed by reversible epigenetic signals such that an epigenetic change influencing one set of neurones will have knock-on effects on many others, and alter our perception and recall of memories. This autobiographical memory (which in all people contains false memories) is a key part of what we call our personal identity, which is made up of a bundle of neural processes, not of any single brain entity.[19] As identical twins have different autobiographical memories, this could explain why even conjoined twins have no problem in separating their identities.

The future

In the classic 1997 SF film *Gattaca* (the title is made up of the four letters of the genetic code), Vincent, played by Ethan Hawke, says: 'I belonged to a new underclass, no longer determined by social status or the colour of your skin. No, we now have discrimination down to a science.' He was conceived and born naturally, which put him at a major disadvantage. His genes, tested routinely at birth, just were not up to scratch. He was correctly predicted to have myopia and a heart defect that would probably kill him at 30.2 years of age. His genes and diseases only allowed him to perform menial tasks. His dream was always to be an astronaut and go to Saturn. He impersonates a genetically perfect alpha male who is paraplegic after a failed suicide bid by using his DNA – and against the odds achieves his dream.

How does the reality stack up? With today's knowledge, how would *Gattaca* play out in the future? We know from our twins that the precise genetic predictions in the film, based on structural DNA, are not going to be possible. In the film, most births are artificial, but we now know that this may present insurmountable epigenetic problems, so as a model of a future society, the *Gattaca* vision may be flawed. However, as in the film, it does seem likely that genetic testing with sequencing will soon be as routinely performed as a blood test is today. It is likely, too, that within ten years gene sequences will be available for the same price as a routine blood screen. The screening test would be performed on the IVF embryo or the pregnant mother's blood, detecting the DNA in a few floating fetal cells. This test would likely be a combination of DNA sequence and epigenetic changes plus a protein and metabolite analysis. The combination of these checks would give the future paediatrician/geneticist a good idea of any potential medical or genetic problems that could be fixed early.

At birth, a repeat of the protein and epigenetic analyses would be carried out, and then repeated at yearly intervals to check development was normal. This will be performed from collections of blood, nail clippings, hair and the lining of the mouth, to check the different cells and tissues. The *Gattaca* infant will be checked for skin, oral and fecal bacteria, to make sure that the correct ratio of healthy bacteria are present. Any abnormalities will be corrected with epigenetic smart drugs and vitamin and bacterial supplements. Functional brain scanning in combination with epigenetic assessments and tailored diets will be used to maximise every baby's potential. Selection of sperm and ovaries that are free of mutations and epigenetic defects will be possible, reducing miscarriages and genetic diseases. The clear difference with the film will be that our understanding of epigenetics and the unpredictable nature of our development will mean that genetics is unlikely to be used to write people off: instead, it might just be possible to use genetics to realise their potential.

Although there is much still to be understood about epigenetics, with what we have learned already we can now rewrite irreversibly at least four key genetic doctrines outlined at the beginning of the book.

The first assumption we have overturned is that our genes are the essence of humans, our blueprint, or the code of life. Because DNA was seen at the centre of all our cells, for a while it was believed to represent the core of life. The gene's over-hyped reputation has been helped by its accepted and infallible use in forensics, as well as through the influence of authors such as Richard Dawkins, who proposed that humans were just robot-like carriers of these self-replicating 'selfish' genes.[20] We have shown how genes, though still important, have lost their privileged and prominent status, particularly as the distinction between nature and nurture disappears.[21]

Genes are certainly star players in the body, but they can't act alone and are one part of a complex team. The cell that hosts them is also a vital member of that team. The cell produces the proteins and enzymes that do the body's work and maintains all the other key cellular mechanisms, including those that turn the genes on and off – from the longer term epigenetic effects such as methylation to the shorter term fine-tuning effects of small RNAs and gene expression. The cell, its genes and its other components could be compared to the orchestra company, with all of its instruments, roadies, and sales and marketing team. Perhaps, though, we have been searching in vain for an elusive conductor of this orchestra – a genetic soul. We are all super-complex organisms formed by interacting networks of cells, their genes, the systems that modify the genes, their expression and the way they work together. It turns out that the rules that govern the orchestra are far from rigid, and the music it produces is anything but predetermined. Just as you can change your lifestyle and the music you make – you can also change your genes.

The second assumption was that genes and heritable genetic destiny can't be changed. But we now know that this is the exception and not the rule. Predictable determination occurs only in rare diseases, as when a single gene mutation in Huntington's disease causes early dementia and death. But even with this disease you can't predict exactly what will happen in terms of disease severity or timing of symptoms. Epigenetics still plays a role in these most simple genetic disorders, and for example clinical trials are ongoing to modify the Huntington gene using the histone-modifying drug SAHA, used in cancer.

The third assumption is that a single environmental event cannot plant a lifelong memory within your cells. It was believed that any changes were wiped clean every time the cell divided. We now know that this memory can occur by epigenetically influencing your genes, which replicate and produce daughter cells with the same epigenetic messages. The most sensitive and influential times for these epigenetic signals are during development before or soon after birth, but they can occur at any age. As we saw with Romanian orphans and abuse victims, the epigenetic signals persist in your cells long after the event.

The final assumption is that the effects of your parents' or grandparents' environments can't be inherited by you. This is the essence of the 'inheritance of acquired characteristics', or 'soft inheritance', proposed by Lamarck and accepted as possible by Darwin. Despite the ridicule this received for most of the last 150 years, we now know that it can occur. We have clearly seen the effects of famine and diet changes across three generations, both in humans and in other mammals, and we now have a mechanism to explain it.

The most important lesson that we've learnt is that you can change your genes, your destiny and that of your children and grandchildren. It really does matter what you do to your body, and importantly what your grandparents did to theirs many years ago. They may have faced stressful situations like famine or sickness that couldn't be avoided, but perhaps you might face life choices like quitting smoking, going vegetarian, or changing your bacterial gut flora. These could influence your life and possibly several generations. We don't know how most of these changes work yet, but if the amazing discoveries from the epigenetic revolution in the last few years are anything to go by, it's going to be an exciting ride. With this knowledge we should be better equipped to shape our destinies, and yet still remain as we were meant to be: identically different. Vive la différence!

ACKNOWLEDGEMENTS

I borrowed many ideas from other clever people, and there are too many books I read and people I quizzed to thank them all, but here are a few of them.

I am indebted to my agents Sophie Lambert and Kevin Conroy Scott for their boundless enthusiasm, advice and encouragement in making this happen. Of the many who read the bad drafts – Victoria Vazquez and Barbara Prainsack read every page and deserve medals. Robyn Fitzgerald, Andrea Burri, Lynn Cherkas and Raj Gill also contributed including help with preparing twin interviews. Lesley Bookbinder, Susan Hochberg and Sophie Spector also helped, as did Emma Walker of the BBC. I had many conversations with helpful colleagues including Jordana Bell, Jonathan Mill, Robert Plomin, Nick Martin, Stephan Beck, Mitch Blair, Roz Kadir, Philip Sambrook, Keith Godfrey, Niloufar Safnia, Bryce from Chatter creek, the guests of the 'Fleur de Lampaul', Jeff Craig, Andy Feinberg and Manel Esteller. I am grateful to KCL and the St Thomas' Charitable trustees for allowing my mini-sabbatical to the CRG Barcelona where Xavier Estivill was a great host, assisted by the inspirational theories of 'the X-Men and Chicas', Elisa doCampo, Monica Coronel and Susana Guzman and the rest of the team. Thanks to Debbie Hart and Victoria who looked after the department in my absence and to my family for allowing my epi-vacation. I'm grateful to the European Research Council for my epigenetics award and the Wellcome Trust and EU for long-term funding. My colleague Wangjun and team at BGI in China have been great collaborators on the EpiTwin project. I would like to thank my editor at Weidenfeld & Nicolson, Bea Hemming, who has been a fantastic support in shaping and improving the drafts.

I have tried to reference as many articles and sources as possible, but many will have been omitted. For this and any factual errors I take the blame. Finally, I'd like to thank all the amazing twin volunteers who willingly and unrewarded give up their stories, time, bits of their bodies and DNA for the advancement of science.

NOTES

Preface

1 Carson, B., *Take the Risk*, Zondervan, 2008

2 The exact number of estimated human genes keep changing – it may be closer to 23,000

3 Carey, N., *The Epigenetics Revolution*, Icon Books, 2011 (a good introduction to the recent science of epigenetics); Epigenetic Epidemiology, *Int J Epidemiology*, Feb 2012, 41; 1: 1–327 (an entire volume on the latest epigenetics from an Epidemiological point of view)

Introduction

1 MacGregor, A. J., et al, *Arthritis Rheum,* January 2000; 43 (1): 30–7

2 Darwin, C., *On the Origin of Species*, 1859, John Murray, London

3 Jablonka, E., and Lamb, M. J., *Evolution in Four Dimensions,* 2005, MIT Press, Cambridge, Mass.

4 www.23andme.com; www.decodeme.com

5 Hindorff, L. A., et al, *Carcinogenesis*, 2011; 32 (7): 945–54; T. A. Manolio, *N Engl J Med*, 2010; 363 (2): 166–76

6 Krause, J., *Nature*, 2007; 449 (7164): 902–4; Scally, A., et al, *Nature*, 7 March 2012 online

7 Klein, R. J., et al, *Science*, 15 Apr 2005; 308 (5720): 385–9

8 Richards, J. B., et al, *Nat Genet*, 2008; 40 (11): 1282–4

9 Visscher, P. M., et al, *Am J Hum Genet*, 2012; 90 (1): 7–24

10 Speliotes, E. K., et al, *Nat Genet*, 2010; 42 (11): 937–48
11 Liu, F., et al, *PLoS Genet*, 6 May 2010; 6 (5): e1000934
12 Doll, R., *Natl Cancer Inst*, June 1981; 66 (6): 1191–1308
13 Scarborough, P., et al, *J Public Health* (Oxford), 11 May 2011
14 Segal, N., *Entwined Lives*, 2000, Plume, New York
15 ibid.; and Segal, N., *Someone Else's Twin*, 2011, Prometheus, New York

Chapter 1 The gene myth

1 Packard, A. S., *Lamarck, The Founder of Evolution: His Life and Work*, Longmans, New York, 1901
2 Lamarck, J.-B., *Philosophie Zoologique*, Dentu, Paris, 1809
3 Packard, A. S., *Lamarck*, Longmans, 1901
4 'Scientist Tells of Success Where Darwin Met Failure', *New York Times*, 3 June 1923, p. 2
5 Noble, G. K., *Nature*, 7 August 1926 209–11
6 *Science*, 19 November 1926; 64 (1664): 493–4
7 Koestler, A., *The Case of the Midwife Toad*, Hutchinson, London, 1971
8 Vargas, A., *J Exp Zool Mol Dev Evol*, 2009
9 Weissman, G., *FASEB*, 2010: 2592–4
10 Joravsky, D., *The Lysenko Affair*, 1986, University of Chicago Press
11 Holliday, R., *Epigenetics*, Apr–Jun 2006; 1 (2): 76–80
12 Bastow, R., et al, *Nature*, 2004; 427: 164–7
13 Cubas, P., et al, *Nature*, 1999; 401: 157–61
14 Bastow, R., et al, *Nature*, 2004; 427: 164–7
15 Schmitz, R. J., et al, *Biochim Biophys Acta*, 2007; 1769 (5–6): 269–75
16 Vandegehuchte, M. B., et al, *Comp Biochem Physiol C Toxicol Pharmacol*, 2009; 50: 343–8

17 Prudic, K. L., et al, *Science*, 7 January 2011; 331 (6013): 73–5

18 Nätt, D., et al, *PLoS ONE*, 2009; 4 (7): e6405. doi: 10. 1371

19 Waterland, R. A., and Jirtle, R. L., *Mol Cell Biol*, 2003; 23: 5293–5300

20 Haig, D., and Graham, C., *Cell*, 1991; 64: 1045–6

21 Nakabayashi, K., et al, *Hum Mol Genet*, 15 August 2011; 20 (16): 3188–97

22 http://www.telegraph.co.uk/health/healthnews/8494966/ Happiness–gene–discovered. html

23 Hammond, C. J., et al, *N Engl J Med*, 2000; 342 (24): 1786–90

24 Spector, T. D., et al, *BMJ*, 1996; 312 (7036): 940–3

25 Liew, S. H., et al, *Twin Res Hum Genet*, June 2005; 8 (3): 198–200

26 MacGregor, A. J., in Spector, T. D., et al, *Advances in Twin and Sib-pair Analyses*, GMMM, 2000, London

27 Parens, E., et al, *Wrestling with Behavioral Genetics*, 2006, Johns Hopkins University Press

28 Kamin, L., *The Science and Politics of IQ*, 1974, Routledge

29 Mackintosh, N. J., *Cyril Burt: Fraud or Framed?*, 1995, Oxford University Press, Oxford; Rose, S., Lewontin, R., Kamin, L., *Not in Our Genes*, Penguin, London, 1990

30 Blinkhorn, S., 'Was Burt Stitched Up?', *Nature*, 1989; 340: 439,

31 Novitski, E., *Genetics*, March 2004; 166: 1133–6

32 Bouchard, T. J., *Science*, 1994; 264 (5166): 1700–01

Chapter 2 'The happiness gene'

1 Provine, R., *Laughter: A Scientific Investigation*, 2001, Viking, Penguin

2 Kendler, K. S., et al, *Psychological Medicine*, 2008; 38: 1475–83

3 Fox, E., *Cogn Emot*, 1 January 2000; 14 (1): 61–92

4 Eysenck, H., *British Journal of Psychology*, 1942; 32: 295–309

5 Vernon, P. A., et al, *Twin Res Hum Genet*, February 2008; 11 (1): 44–7

6 Ibid.

7 Buijzen, M., et al, *Journal of Applied Developmental Psychology*, 2003; 24: 437–56

8 Giddens, J., et al, *Personality and Individual Differences*, 2009; 46 (4): 428–31

9 Sharot, T., *The Optimism Bias*, Pantheon, 2011

10 Diener, E., *American Psychologist,* 2000; 55: 34–43

11 Helliwell, J. F., *Economic Modelling*, 2003; 20 (2): 331–60

12 http://www.guardian.co.uk/global-development/2011/jan/05/nigeria-happiness-polls?intcmp=239

13 Lykken, D., and Tellegen, A., *Psychological Science*, 1996; 7 (3): 186–9

14 Weiss, A., et al, *Psychol Sci*, March 2008; 19 (3): 205–10

15 McGue, M., *Psychol Aging*, March 1993; 8 (1): 72–80

16 Plomin, R., et al, *Personality and Individual Differences,* 1992; 13: 921–30

17 Chida, Y., et al, *Psychosomatic Medicine*, 2008; 70 (7): 741–56

18 Rasmussen, H., et al, *Annals of Behavioral Medicine*, 2009; 37 (3): 239–56

19 Puri, M., *Journal of Financial Economics*, 2007; 86: 71–99

20 Friedman, H. S., and Martin, L. R., *The Longevity Project*, Hay House, London, 2011

21 Collins, J., *Good to Great*, 2001, HarperCollins, New York

22 Ryan, L., et al, *Br J Psychiatry*, June 2006; 188: 560–6

23 Laje, G., *Prog Neuropsychopharmacol Biol Psychiatry*, 2011; 35 (7): 1553–7

24 Sullivan, P. F., et al, *Am J Psychiatry*, 2000; 157 (10): 1552–62

25 Lesch, K. P., et al, *Science*, 1996; 274: 1527–31

26 Hariri, A. R., et al, *Science*, 2002; 297: 400–04; A. Caspi et al, *Am J Psychiatry*, 2010; 167 (5): 509–27

27 Karg, K., et al, *Arch Gen Psychiatry,* 2011; 68 (5): 444–54

28 Kinnally, E. L., et al, *Genes Brain Behav*, 2010; (6): 575–82

29 Fox, E., et al, *Proc Biol Sci*, 22 May 2009; 276 (1663): 1747–51

30 Dweck, C., *Mindset: The New Psychology of Success*, Ballantine Books, 2008; Dweck, C., *Mindset: How you can fufill your potential*, Constable, 2012

31 Steele, C. M., et al, *J Pers Soc Psychol*, 1995; 69 (5): 797–811

32 Seligman, M. E., et al, *Am Psychol*, 2006; 61 (8): 774–88; Seligman, M. E., (1991), *Learned Optimism: How to Change Your Mind and Your Life*, Penguin, New York, 1998; Seligman, M. E., *Flourish*, Random House, 2011

33 Lutz, A., et al, *PNAS,* 2004; 101 (46): 16369–73

34 Ott, U., et al, *Studies in Neuroscience, Consciousness and Spirituality*, 2011; 1: 119–28

35 Fowler, J. H., et al, *BMJ,* 2008; 337: a2338

Chapter 3 'The talent gene'

1 http://www.cygenedirect.com/browse-10873/Optimum-Athletic-Performance-DNA-Analysis.html

2 www.atlasgene.com

3 Montgomery, H. E., et al, *Nature*, 1998; 393: 221–2

4 Montgomery, H. E., et al, *Circulation*, 1997; 96: 741–7

5 Yang, N., et al, *Am J Hum Genet*, 2003; 73 (3): 627–31

6 MacArthur, D. G., et al, *Exerc Sport Sci Rev*, January 2007; 35 (1): 30–4

7 Puthucheary, Z., et al, *Sports Med*, 1 Jun 2011; 41 (6): 433–48; Döring, F., et al., *Journal of Sports Sciences*, 28, 12: 1355–9; Hagberg, J. M., *Med Sci Sports Exerc*, May 2011; 43 (5): 743–52

8 Theusch, E., et al, *Twin Res Hum Genet*, April 2011; 14 (2): 173–8

9 Gladwell, M., *Outliers*, Penguin, 2009; Syed, M., *Bounce*, Fourth Estate, 2010; Colvin, G., *Talent Is Overrated*, Nicholas Brealey, 2008; Shenk, D., *The Genius in All of Us*, Icon, 2010; Coyle, D., *The Talent Code*, Arrow Books, 2010

10 Ericsson, K. A., et al, *Harvard Business Review*, July–August 2007: 115–21

11 Ericsson, K. A., et al, *Psychological Review*, 1993; 100: 363–406

12 Pott, J., 'The Triumph of Genius: Celebrating Mozart', *Books and Culture*, November–December 2006

13 Syed, M., *Bounce*, Fourth Estate, London, 2010; Epstein, D., 'Sports Genes' in *Sports Illustrated*, May 2010

14 Bramble, D. M., et al, *Nature*, 2004; 432: 345–52

15 Maguire, E. A., et al, *The Journal of Neuroscience*, 1997; 17 (18): 7103–10; Maguire, E. A., et al, *Science*, 280 (5365): 921–4

16 Woollett, K., et al, *Philos Trans R Soc Lond B Biol Sci*, 2009; 364 (1522): 1407–16

17 Ericcson, K. A., et al, *Science*, 1980; 208 (4448): 1181–2

18 Forbes, C., *The Polgar Sisters: Training or Genius?*, Henry Holt, New York, 1992

19 Galton, F., *Hereditary Genius*, Macmillan, London, 1869

20 Stubbe, J. H., et al, *PLoS ONE*, 20 Dec 2006; 1: e22

21 Bouchard, C., et al, *J Appl Physiol*, May 2011; 110: 1160–70;

North, K. N., et al, *Nature Genetics*, 21: 353–4; Bouchard, C., et al, *Medicine and Science in Sports and Exercise*,1998; 30: 252–8

22 De Moor, M. H., et al, *Twin Res Hum Genet*, Dec 2007; 10 (6): 812–20

23 Arden, N. K. and Spector, T. D., *J Bone and Mineral Research*, 1997; 12: 2076–81; Vink, J. M., *Twin Res Hum Genet*, Feb 2011; 14 (1): 25–34

24 Drayna, D., et al, *Science*, 2001; 291 (5510): 1969–72

25 Peretz. I., et al, *Am J Hum Genet*, 2007; 81: 582–8

26 Ukkola-Vuoti, L., et al, *J Hum Genet*, Apr 2011; 56 (4): 324–9

27 Loui, P., et al, *J Cogn Neurosci*, 2011; 23 (4): 1015–26

28 Miyazaki, K., *Percept Psychophys*, 1988; 44 (6): 501–12

29 Theusch, E., et al, *Twin Res Hum Genet*, Apr 2011; 14 (2): 173–8

30 Nisbett, R. E., et al, *Am Psychol*, 2 Jan 2012 (epub)

31 Duckworth, A. L., et al, *Proc Natl Acad Sci USA*, 10 May 2011; 108 (19): 7716–20

32 Dr James Thompson, of UCL London, said: 'Life is an IQ test and a personality test and an IQ result contains elements of both (but mostly intelligence). If an IQ test doesn't motivate someone then that is a good predictor in itself.'

33 Scarr, S., *Child Development*, 1966; 37 (3): 663–73

34 Terruzzi, I., et al, *Physiol Genomics*, 2011; 43 (16): 965–73

35 Zhang, F. F., et al, *Epigenetics*, 1 March 2011; 6 (3): 293–9; Yuasa, Y., et al, *Int J Cancer*, 2009; 124 (11): 2677–82; Nakajima, K., et al, *Int J Sports Med*, 2010; 31 (9): 671–5

36 Collins, A., et al, *PLoS ONE*, 2009; 4 (1): e4330

37 Lightfoot, J. T., et al, *Physiological Genomics*, 2004; 19: 270–6

38 Brigati, C., et al, *Frontiers in Genetics*, 12 January 2012 (online)
39 Day, J. J., et al, *Neurobiol Learn Mem*, July 2011; 96 (1): 2–12
40 Baumeister, R. F., and Tiereny, J., *Willpower*, Allen Lane, 2011
41 Sharp, N. C., *Epigenetics*, 2011; 6 (3): 293–9
42 Day, J.J., et al, *Neuropsychopharmacology*, Jan 2012; 37 (1): 247–60
43 Mill, J., *Neurobiol Aging*, Jul 2011; 32 (7): 1188–91

Chapter 4 'The God gene'

1 Leege, D. C., et al, *Rediscovering the Religious Factor in American Politics*, NY, 1993, M. E. Sharpe
2 Diener, E., *Journal of Personality and Social Psychology*, 2011; 101 (6): 1278–90
3 Hebrews, 11:1, KJV
4 Bishop, G. F., et al, *Americans' Scientific Knowledge*, 2010; National Center for Science Education (NCSE)
5 Balmer, R., www.washingtonmonthly.com/features/2007/0704.balmer.html
6 Koenig, J., and Bouchard, T., in McNamara, *Where God and Science Meet*, Praeger, 2006
7 Park, A., et al, *Religion in Britain and the US: British Social Attitudes 2009–10*, Sage, 2010
8 Kirk, K. M., et al, *Twin Res*, June 1999; 2 (2): 81–7
9 Vance, T., et al, *J Nerv Ment Dis*, 2010; 198 (10): 755–61
10 Koenig, L. B., et al, *J Pers*, 2005; 73 (2): 471–88
11 Collins, F., *The Language of God*, Simon and Schuster, London, 2007
12 Barlow, D., *The Autobiography of Charles Darwin, 1809–1882*, Collins, London, 1958, pp. 92–4

13 Culotta, E., *Science*, 6 November 2009; 326 (5954): 784–7

14 Dennett, D., *Breaking the Spell*, Penguin, London, 2007

15 Nicholson, A., et al, *Soc Sci Med*, August 2009; 69 (4): 519–28

16 Lin, K., et al, *Epilepsy Behav*, February 2009; 14 (2): 400–3

17 Kapogiannis, D., et al, *PLoS ONE*, 28 September 2009; 4 (9): e7180

18 Ackland, G., *PNAS*, 2007; 8714–19

19 Ramsey, S., et al, *J Sex Med*, 2009; 6 (8): 2102–10

20 Herbenick, D., et al, *J Sex Med*, 2010; 7 (10): 3322–30

21 Schick, V. R., *J Sex Res*, 2011; 48 (1): 74–81

22 Hirschle, J., *J Scientific Study of Religion*, 2010; 673–87

23 http://www.time.com/time/world/article/0,8599,1942665,00.html#ixzz1D0zwbDSv

24 McAndrew, S., 'Religious Faith and contemporary attitudes', in Park et al, *British Social Attitudes, 2009–2010*, Sage, 2010

25 Greely, J., *J Scientific Study of Religion*, 1994, 253–72

26 Greksa, L. P., et al, *Ann Hum Biol*, 2002; 29 (2): 192–201

27 Rowthorn, R., *Proc Biol Sci*, 22 August 2011; 278 (1717): 2519–27

28 Park et al, *British Social Attitudes, 2009–10*; http://www.timesonline.co.uk/tol/news/uk/article5621482.ece

29 Hamer, D., *The God Gene*, Doubleday, New York, 2004

30 Koenig, J., et al in McNamara, P., *Where God and Science Meet*, Praeger, 2006, 31–50

31 Irwin, H. J., *The Psychology of Paranormal Belief: A Researcher's Handbook*, Univ Herts Press, 2009

32 Ridolfo, H., et al, *Current Research in Social Psychology*, 2010; 15: 33–41

33 Bering, J., *The God Instinct*, Nicholas Brealey, London, 2011

Chapter 5 'The parenting gene'

1 Hardyment, C., *Dream Babies: Childcare Advice from John Locke to Gina Ford*, Frances Lincoln, London, 2007
2 Aries, P., *Centuries of Childhood: A Social History of Family Life*, Random House, 1965
3 Caplan, B., *Selfish Reasons to Have More Kids*, Basic Books, New York, 2011
4 Chua, A., *Battle Hymn of the Tiger Mother*, Bloomsbury, 2011
5 Spock, B., *Dr. Spock's Baby and Childcare*, Seventh Edition, Pocket, 1998
6 Ford, G., *The Contented Little Baby*, NAL, New York, 2001
7 Sears, M., and Sears, W., *The Complete Book of Christian Parenting and Child Care*, Broadman and Holman, Nashville, 1997
8 James, O., *How Not to F*** Them Up*, Vermillion, 2011; Leach, P., *Your Baby and Child*, Knopf, 2010
9 Bettelheim, B., *A Good Enough Parent: A Book on Child-Rearing*, Vintage, 1988
10 Harris, J. R., *The Nurture Assumption*, Free Press, 2009
11 Dunn, J., and Plomin, R., *Separate Lives: Why siblings are so different*, Basic Books, 1990
12 Turkheimer, E., et al, *Psychol Bull*, January 2000; 126 (1): 78–108
13 Neiderhiser, J. M., et al, *Twin Res Hum Genet*, 2007; 10 (1): 74–83
14 Reiss, D., et al, *Am J Psychiatry*, 1991; 148 (3): 283–91
15 Reiss, D., *Acta Psychol Sinica*, 2000; 40 (10): 1099–1105
16 Schulz-Heik, R. J., *Behav Genet*, May 2010; 40 (3): 338–48
17 Pike, A., et al, *Dev Psychol*, 1996; 32: 590–603
18 Plomin, R., et al, *Int J Epidemiol*, 2011, 40 (3): 563–82

19 Asbury, K., et al, *Twin Research and Human Genetics*, 2008; 11: 586–95
20 Harlow, H. F., et al, *Proc Natl Acad Sci USA*, July 1965; 54 (1): 90–7
21 Van Cleve, T. C., *Emperor Frederick II of Hohenstaufen: Immutator Mundi*, OUP, 1972
22 Rutter, M. L., et al, *The British Journal of Psychiatry*, 2001; 179: 97–103
23 Rutter, M. L., et al, *Dev Psychol*, 2004; 40: 81–94
24 Schulz-Heik, R. J., et al, *Behavior Genetics*, 2010; 40: 338–48
25 Stevens, S., et al, *Am J Med Genet Part B*, 2009; 150B: 753–61
26 Cirulli, F., et al, *Neuroscience and Biobehavioral Reviews*, 2009; 33: 573–85
27 Weaver, I. C., et al, *Nature Neuroscience*, 2004; 7: 847–54
28 McGowan, P. O., et al, *PLoS ONE*, 28 Feb 2011; 6 (2): e14739
29 Shahrokh, D. K., et al, *Endocrinology*, 2010; 151: 2276–86
30 Febo, M., et al, *J Neuroscience*, 2005; 25: 11637–44
31 Cameron, N. M., et al, *J Neuroendocrinol*, 2008; 20: 795–801
32 Champagne, F. A., et al, *J Behavioral Neuroscience*, 2007; 121 (6): 1353–63; Champagne, F. A., et al, *Biol Psych*, 2006; 59: 1227–35
33 Bowes, L., et al, *J Child Psychol Psychiatry*, 2010; 51 (7): 809–17
34 Ball, H. A., et al, *J Child Psychol Psychiatry*, 2008; 49 (1): 104–12
35 Burt, S. A., et al, *Psychological Bulletin*, 2009; 135, 608–37
36 Eaves, L. J., et al, *Behav Genet*, 2010; 40 (4): 425–37

Chapter 6 'Bad genes'

1 Gorey, K. M., et al, *Child Abuse and Neglect*, 1997; 21 (4): 391– 8; Finkelhor, D., *The Future of Children*, 1994; 4 (2): 31–53

2 Baker, A. W., et al, *Child Abuse and Neglect*, 1985; 9 (4): 457–67; Rind, B., et al, *Psychological Bulletin*, 1998; 124 (1): 22–53; 'Child Maltreatment 2005'. Administration on Children and Families. US Department of Health and Human Services, 26 December 2007

3 Gorey, K. M., et al, *Child Abuse and Neglect*, 1997; 21 (4): 391–8

4 Lascaratos, J., et al, *Child Abuse and Neglect*, 2000; 24, 8: 1085–1090

5 Rezmovic, E. L., et al, 1996, http://www.ncjrs.gov/App/Publications/abstract.aspx?ID=165912

6 Ball, H. A., et al, *J Child Psychol Psychiatry*, 2008; 49 (1): 104–12

7 Dinwiddie, S., et al, *Psychological Medicine*, 2000; 30: 41–52

8 Schulz-Heik, R. J., et al, *Behav Genet*, 2009; 39: 265–76

9 Kendler, K. S., et al, *Archives of General Psychiatry*, 2000; 57 (10): 953–9

10 Weich, S., et al, *The British Journal of Psychiatry*, 2009, 194: 392–8; Hill, J., et al, *The British Journal of Psychiatry*, 2001, 179: 104–9

11 Kendler, K. S., et al, *Archives of General Psychiatry*, 2000; 57 (10): 953–9

12 Rind, B., et al, *Journal of Sex Research*, 1997; 34 (3): 237–55

13 Dallam, S. J., et al, *Psychological Bulletin*, 2001; 127 (6): 715–33

14 US Congress (1999). 'Whereas no segment of our society is

more critical to the future of human survival than our children' (PDF). 106th Congress, Resolution 107

15 Ulrich, H. A., 'Replication of the Meta-analytic Examination of Child Sexual Abuse by Rind, Tromovitch, and Bauserman (1998)' PhD thesis, 2004 http://worldcat.org/oclc/55532119

16 Kar, N., et al, *Neuropsychiatr Dis Treat*, 2011; 7: 167–81

17 LeardMann, C. A., et al, *BMJ*, 2009; 338: b1273

18 Loftus, E. F., and Pickrell, J. E., *Psychiatric Annals*, 1995; 25: 720–5

19 Spitzer, R. L., et al, *Journal of Anxiety Disorders*, 2007; 21 (2): 233–41

20 Rosen, G. M., et al, *Clinical Psychology Review*, 2008; 28 (5): 837–68

21 Erbes, C., et al, *Military Medicine*, 2007; 172: 359–63; Hoge, C. W., et al, *New England Journal of Medicine*, 2004; 351: 13–22

22 Heath, A. C., et al, *Psychol Med*, 1997; 27 (6): 1381–96

23 Machado, H., and Prainsack, B., *Tracing Technologies: Prisoners' Views in the Era of CSI*, Ashgate, 2012. http://www.cambridge.org/gb/knowledge/isbn/item5562820/

24 Vaughn, M. G., et al, *Criminal Justice and Behavior*, 2009; 36 (11): 1113–24; DeLisi, M., et al, *Criminal Justice and Behavior*, 2009; 36 (11): 1241–52

25 Baron-Cohen, S., *Zero Degrees of Empathy*, Allen Lane, London, 2011

26 Zanarini, M. C., *Psychiatric Clinics of North America*, 2000; 23: 89–101; Bryer, J., *Am J Psych*, 1987; 1426–30; Caspi, A., et al, *Science*, 2002; 297 (5582): 851–4

27 Francis, K. J., et al, *Child Abuse Negl*, 2008; 32 (12): 1127–37

28 Centerwall, B. S., et al, *Compr Psychiatry*, 1989; 30 (5): 442–6

29 Lyons, M. J., et al, *Arch Gen Psychiatry*, 1995; 52 (11): 906–15

30 Lichtenstein, P., et al, *Twin Res Hum Genet*, 2007; 10 (1): 67–73

31 Lyons, M. J., 'A Twin Study of Self-Reported Criminal Behaviour' in Ciba Foundation Symposium 194 – Genetics of Criminal and Antisocial Behaviour (eds G. R. Bock and J. A. Goode), John Wiley and Sons, 2007

32 Christiansen, K. O., *Int J of Criminology and Penology*, 1973; 1, 1: 31–45; Lyons, M. J., et al, *Arch Gen Psychiatry*, November 1995; 52 (11): 906–15

33 Rhee, S. H., and Waldman, I. D., *Psychol Bull*, 2002; 128: 490–529

34 Bohman, M., 'Predisposition to Criminality: Swedish Adoption Studies in Retrospect', in *Ciba Foundation Symposium 194 – Genetics of Criminal and Antisocial Behaviour* (eds G. R. Bock and J. A. Goode), John Wiley and Sons, 2007

35 Brambilla, P., *Psychiatry Res*, 2004; 131 (2): 125–33

36 Rijsdijsk, F. V., et al, *Arch Gen Psychiatry*, 2010; 67 (4): 406–13

37 McGowan, P. O., et al, *PLoS ONE*, 28 February 2011; 6 (2): e14739

38 Ernst, C., et al, *Arch Gen Psychiatry*, 2009; 66 (1): 22–32

39 McGowan, P. O., et al, *Neurobiol Dis*, 2010; 39 (1): 66–72

40 McGowan, P. O., et al, *Nat Neurosci*, 2009; 12: 342–48

41 Ponseti, J., et al, *Arch Gen Psychiatry*, 2012; 69 (2): 187–94

42 Maes, H. H., et al, *Twin Res Hum Genet*, 2007; 10 (1): 136–50

43 Essex, M. J., et al, *Child Dev*, 2011, doi. 0.1111

44 Tarantino, L. M., et al, *Front Psychiatry*, 2011; 2: 44

45 Allen, G., and Duncan-Smith, I., *Early Intervention: Good Parents, Great Kids, Better Citizens*, The Centre for Social Jus-

tice and the SMITH Institute, 2008

46 'Fair Society, Healthy Lives', *The Marmot Review*, The Dept of Health, 2010

47 Olds, D., *Prevention Science*, 2002; 3: 153–72; Olds, D., et al, *JAMA*, 1998; 278: 637–43

48 Kendrick., D, et al, *Evidence-Based Child Health*, 2009; 4: 1005–39

49 Dretzke, J., et al, *Child and Adolescent Psychiatry and Mental Health*, 2009; 3: 7

50 McMunn, A. M., *Social Science and Medicine*, 2001; 53: 423–40

51 Belsky, J., and Beaver, K. M., *J Child Psychol Psychiatry*, 2011; 52 (5): 619–26

Chapter 7 'The mortality gene'

1 Zdravkovic, S., et al, *Journal of Internal Medicine*, 2002; 252: 247–54

2 Noble, D., *The Music of Life*, Oxford University Press, 2006

3 Barker, D. J. P., et al, *The Lancet*, 1989; 334: p. 577–80

4 Dikötter, F., *Mao's Great Famine: The History of China's Most Devastating Catastrophe, 1958–62*, Walker & Company, 2010

5 Bygren, L. O., et al, *Acta Biotheoretica*, 2001; 49: 53–9; Kaati, G., et al, *Eur J Hum Genet*, 2002; 10: 682–8

6 Kaati, G., et al, *Eur J Hum Genet*, 2007; 15: 784–90

7 Pembrey, M. E., et al, *Eur J Hum Genet*, 2006; 14: 159–166

8 Chen, T. H., et al, *Am J Clin Nutr*, 2006; 83 (3): 688–92; Lin, W. Y., et al, *Obesity*, 2009; 17 (6): 1247 –54

9 Boucher, B. J., et al, *Diabetologia*, 1994; 37: 49–55

10 Heijmans, B. T. J., et al, *Proc Natl Acad Sci USA*, 2008; 105: 17046–9

11 Hughes, L. A., et al, *Int J Epidemiol*, 2010; 39 (5): 1333–44;

epub, 28 April 2010

12 Cherkas, L. F., et al, *Aging Cell*, 2006; 5 (5): 361–5

13 Marmot, M.G., et al, *The Lancet*, 1991; 337: 1387–93

14 Sapolsky, R., *Why Zebras Don't Get Ulcers: An Updated Guide To Stress, Stress Related Diseases, and Coping*, Scientific American Library, 1998

15 Tung, J., et al, *PNAS*, epub, 9 April 2012

16 Valdes, A. M., et al, *The Lancet*, 2005; 366 (9486): 662–4

17 Epel, E. S., et al, *Hormones*, 2009; 8 (1): 7–22

18 Marmot, M. G., et al, *JAMA*, 2009; 301 (11): 1169–71

Chapter 8 'The fat gene'

1 Swami, V., *J of Sex Research*, 2009; 46 (1): 89–96

2 CDC 2004 and NHANES surveys

3 Scuteri, A., et al, *PLoS Genet*, July 2007; 3 (7): e115; Willer, C. J., et al, *Nat Genet*, January 2009; 41 (1): 25–34

4 Church, C., et al, *Nat Genet*, 2010; 42 (12): 1086; Cecil, J. E., et al, *N Engl J Med*, 2008; 359 (24): 2558–66; Timpson, N. J., et al, *Am J Clin Nutr*, 2008; 88 (4): 971–8; Olszewski, P. K., et al, *Biochem Biophys Res Commun*, 2011; 408 (3): 422–6; McTaggert, J. S., *PLoS ONE*, 2011; 6 (11): e27968

5 Falchi, M., et al, 2012 (submitted)

6 Kilpeläinen, T. O., et al, *PLoS Med*, 2011; 8 (11): e1001116

7 Leskelä, P., et al, *Metabolism*, Feb 2009; 58 (2): 174–9; Zac-Varghese, S., et al, *Discov Med*, 2010; 10 (55): 543–52

8 Romero, A., et al, *Eur J Endocrinol*, 2010; 163 (1): 1–8

9 Kilpeläinen, T. O., et al, *Nature Genetics*, 2011; 43: 753–60

10 Speliotes, E. K., et al, *Nat Genet*, November 2010; 42 (11): 937–48; Heid, I. M., et al, *Nat Genet*, November 2010; 42 (11): 949–60

11 Samaras, K., et al, *J Clin Endocrinol Metab*, March 1997; 82 (3): 781–5
12 Small, K. S. et al, *Nat Genet*, June 2011; 43 (6): 561–4
13 Almén, M. S., et al, *Genomics*, 2012; 99 (3): 132–7
14 Lack, G., 'Food Allergy', *N Engl J Med*, 2008; 359: 1252–60
15 Itan, Y., et al, *BMC Evol Biol*, 9 February 2010; 10: 36
16 Teucher, B., et al, *Twin Res Hum Genet*, October 2007; 10 (5): 734–48
17 Schumann, G., et al, *Proc Natl Acad Sci USA*, 2011; 108 (17): 7119–24
18 Fraser, G. E., et al, *Arch Intern Med*, 2001; 161: 1645–52
19 Waterland, R. A., and Jirtle, R. L., *Mol Cell Biol*, 2003; 23 (15): 5293–5300; Whitelaw, N. C., and Whitelaw, E., *Curr Opin Genet Dev*, 2008; 18: 273–9; Jirtle, R. L., and Skinner, M. K., *Nat Rev Genet*, 2007; 8: 253–62
20 MRC Vitamin Study Research Group, *Lancet*, 1991; 338: 131–7
21 Zambrano, E., et al, *J Physiol*, 1 July 2005; 566 (1): 225–36
22 Cesani, M. F., et al, *Cells Tissues Organs*, 2003; 174: 129–35
23 Ng, S. F., *Nature*, 2010; 467: 963–66
24 Carone, B. R., et al, *Cell*, 2010; 143 (7): 1084–96
25 Shoda, Y., et al, *Developmental Psychology*, 1990; 26 (6): 978–986
26 Mischel, W., et al, 'Delay of gratification in children', *Science*, 1989; 244: 933–8
27 Norbury, T. A., et al, *Brain*, 2007; 130 (Pt 11): 3041–9
28 Baumeister, R. F., and Tiernet, J., *Willpower: Rediscovering Our Greatest Strength*, Penguin, London, 2012
29 www.mint.com, www.rescuetime.com, www.lifehacker.com
30 Woolgar, C. M., et al, *Food in Medieval England: Diet and Nutrition*, OUP, 2006; Scully, T., *The Art of Cookery in the Middle Ages*, The Boydell Press, Woodbridge, 1995

31 Mozaffarian, D., et al, *N Engl J Med*, 2011; 364: 2392–404
32 Cereda, E., et al, *Clin Nutr*, 2011; 30 (6): 718–23; Weinsier, R. L., et al, *Am J Clin Nutr*, 2000; 72: 1088e94.
33 Sumithran, P., et al, *N Engl J Med*, 2011; 365 (17): 1597–1604
34 Levine, J. A., *Am J Physiol Endocrinol Metab*, 2004; 286 (5): E675–85
35 Bouchard, C., et al, *N Engl J Med*, 1990; 322 (21): 1477–82
36 Hainer, V., et al, *Int J Obes Relat Metab Disord*, August 2000; 24 (8): 1051–7
37 Bouchard, L., et al, *Am J Clin Nutr*, 2010; 91 (2): 309–20
38 Godfrey, K. M., et al, *Diabetes*, May 2011; 60 (5): 1528–34
39 http://store.livingthecrway.com/
40 Tissenbaum, H. A., et al, *Nature*, 2001; 410: 227–30
41 Guarente, L., *N Engl J Med*, 2011; 364: 2235–44
42 Howitz, K. T., et al, *Nature*, 2003; 425 (6954): 191–6
43 Burnett, C., et al, *Nature*, 21 September 2011; 477 (7365): 482–5; J. Van Raamsdonk, i PNAS, 10 April 2012, 109 (15), 5785–90
44 Kanfi, Y., et al, *Nature*, 22 February 2012, 438 (7388): 218–21
45 Markert, C. D., et al, *J Med Food*, 2010; 13 (5): 1081–5
46 Hamadeh, M. J., et al, *Muscle Nerve*, 2005; 31 (2): 214–20
47 Valdes, A. M., et al, *Lancet*, 2005; 366 (9486): 662–4
48 Baur, J. A., *Mech Ageing Dev*, April 2010; 131 (4): 261–9
49 Park, S. J., et al, *Cell*, February 2012; 148 (3): 421–33
50 Corder, R., *Nature*, 2006; 444 (7119): 566
51 Goldberg, D. M., et al, *Clinical Biochemistry*, 2003; 36 (1): 79–87
52 Wing, R. R., et al, *Obesity Res*, September 2004; 12 (9): 1426–34

53 Goldacre, B., *Bad Science*, HarperCollins, London, 2009 (www.badscience.net)

54 Bjelakovic, G., et al, *JAMA*, 2007; 297: 842–857; Bjelakovic, G., et al, *Cochrane Database Syst Rev*, 2008; 2: Art No: CD007176

55 Omenn, G. S., et al, *N Engl J Med*, 1996; 334: 1150–5; The Alpha-Tocopherol, Beta Carotene Cancer Prevention Study Group, *N Engl J Med*, 1994; 330: 1029–35

56 Goldacre, B., *Bad Science*, HarperCollins, London, 2009 (www.badscience.net)

57 Bolland, M. J., et al, *BMJ*, 19 April 2011; 342; Bolland, M. J., et al, *Am J Clin Nutr*, 2011; 94 (4): 1144–9

Chapter 9 'The cancer gene'

1 MacGregor, A. J., et al, *Arthritis Rheum*, 2004; 51 (2): 160–7

2 Curtis, C., et al, *Nature*, 18 April 2012, online

3 Antoniou, A., et al, *Am J Hum Genet*, 2003; 72 (5): 1117–30

4 Global Cancer Facts and Figures 2007. Atlanta, GA: American Cancer Society, 2007

5 Porter, P. L., *Salud Publica Mex*, 2009; 51 (Suppl 2): s141–6

6 Gladwell, M., 'John Rock's Error', *The New Yorker*, retrieved 2007

7 Reeves, G. K., et al, *BMJ*, 2007; 335 (7630): 1134

8 Xue, F, et al, *Lancet Oncology*, 2007; 8: 1088–1100; *Lancet*, 1996; 347: 1713–27

9 Boyd, N. F., et al, *Br J Cancer*, 2003; 89 (9); 1672–85; Thiebaut, A. C., et al, *J Natl Cancer Inst*, 2007; 99: 451–62; Prentice, R. L., et al, *JAMA*, 2006; 295: 629–42

10 Velentzis, L. S., et al, *Br J Cancer*, 2009; 100 (9): 1492–8

11 Jenkins, S., et al, *J Steroid Biochem Mol Biol*, 23 June 2011 (epub)

12 Hansen, K. D., et al, *Nat Genet*, 2011; 43 (8): 768–75
13 Fang, F., et al, *Sci Transl Med*, 23 March 2011; 3 (75): 75ra25
14 Boumber, Y., and Issa, J. P., *Oncology*, March 2011; 25 (3): 220–6, 228
15 Raynal, N. J., et al, *Cancer Res*, 4 January 2012 (epub)
16 Wu, H., et al, *Nature*, 19 May 2011; 473 (7347): 389–93
17 Hansen, K. D., et al, *Nat Genet*, 2011; 43 (8): 768–75
18 www.epigenomics.com; Ladabaum, U., et al, *Digestive Disease Week*, 7–10 May 2011, Chicago, IL (Abstract #220)
19 Heyn, H., et al, *PLoS Genetics*, submitted 2012
20 Assis, S., 'Transgenerational inheritance of breast cancer', 2011, Abstract, March, Keystone, p. 73
21 Jenkins, S., et al, *J Steroid Biochem Mol Biol*, 23 June 2011 (epub)
22 McLachlan, J. A., et al, *Best Pract Res Clin Endocrinol Metab*, March 2006; 20 (1): 63–75
23 Treffert, D. A., *Philos Trans R Soc Lond B Biol Sci*, 2009; 364 (1522): 1351–7
24 Mahler, M. S., Furer, M., Settlage, S. F., 'Severe emotional disturbances in childhood: psychosis', in Arieti, S., (ed), *American Handbook of Psychiatry*, Basic Books, 1959, pp. 816–39
25 Kim, Y. S., et al, *Am J Psychiatry*, 9 May 2011 (epub)
26 Hallmayer, J., et al, *Arch Gen Psychiatry*, 4 July 2011 (epub)
27 Wang, K., et al, *Nature*, 28 May 2009; 459 (7246): 528–33
28 Berkel, S., et al, *Nature Genetics*, 2010; 42: 489–91
29 Slate, M. W., *Nature Genetics*, 2010; 42: 478–9
30 Rizzolatti, G., et al, *Exp Brain Res*, 2010; 200: 223–37
31 Altschuler, E. L., et al. [abstract 67. 23]. 30th annual mtg soc for neuroscience, New Orleans, 4–9 November 2000
32 Ramachandran, V. S., and Oberman, L. M., 'Broken

Mirrors: A Theory of Autism', *Sci Am*, 2006; 295: 62–9

33 Ramachandran, V. S., *The Tell-Tale Brain*, Random House, London, 2011

34 Ramachandran, V. S., and Seckel, E. L., *Med Hypotheses*, January 2011; 76 (1): 150–1

35 Healey, K., *Two Brothers, Worlds Apart*, SAAS, 2011

36 Hertz-Picciotto, I., and Delwiche, L., *Epidemiology*, 2009; 20 (1): 84–90; Grether, J. K., et al, *J Autism Dev Disord*, 2009; 39 (10): 1412–19

37 Newschaffer, C. J., et al, *Pediatrics,* 2005; 115 No. 3 e277–e282

38 Croen, L. A., et al, *Arch Gen Psychiatry*, 2011; 68 (11): 1104–12

39 Wakefield, A. J., et al, *The Lancet*, 17 April 2004; 363 (9417): 1327–8

40 Baron-Cohen, S., *PLoS Biol*, June 2011; 9 (6): e1001081

41 Skuse, D. H., *Pediatric Research*, 2000; 47: Issue 1, p. 9

42 Stoltenberg, C., *Epidemiology*, 2011; 22: Issue 4, pp. 489–90

43 Nguyen, A., et al, *FASEB*, August 2010; 24 (8): 3036–51

44 Golding, J., et al, *PLoS ONE*, 2010; 5 (4): e9939

45 Foldi, C. J., et al, *Front Behav Neurosci*, 2011; 5: 32; (epub 23 June 2011)

46 Zerbo, O., et al, *Epidemiology*, 2011; 22: 469–75

47 http://www.trccorp.com/faq_root_disease6.php

48 Schmidt, R. J., et al, *Epidemiology*, 2011; 22: 476–85

49 Mehler et al, *Brain Research Reviews*, 2009; 59 (2): 388

50 Wolstenholme, J. T., et al, *Horm Behav*, March 2011; 59 (3): 296–305

51 Dolinoy, D. C., et al, PNAS, 2007; 104 (32): 13056

52 Palanza, P., et al, *Environ Res*, 2008; 108: 150–7

53 Nakagami, A., et al, *Psychoneuroendocrinology*, September

2009; 34 (8): 1189–97

54 Xu, X. H., et al, *Hormone Behaviour*, July 2010; 58 (2): 326–33

55 Jasarevic, E., et al, *Proc Natl Acad Sci USA*, 12 July 2011; 108 (28): 11715–20

56 Salian, S., et al, *Life Sci*, 2009; 85 (1–2): 11–18

57 Vallero, R., and LaSalle, J. M., 'Epigenetics at the interface of Genetic and Environmental risk factors for ASD,' Abstract, Keystone, March 2011, p. 60

58 Anway, M. D., et al, *Endocrinology*, 2006; 147: 5515–23

59 Vandenberg, L. N., et al, *Reprod Toxicol*, August–September 2007; 24 (2): 139–77

60 Lang, I. A., et al, *JAMA*, 2008; 300 (11): 1303–10

61 Melzer, D., et al, *PLoS ONE*, 2010; 5 (1): e8673

62 Clayton, E. M., et al, *Environ Health Perspect*, 2011; 119: 390–6

63 Ugiura-ogasawara, M., et al, *Human reproduction*, 2005; 20 (8): 2325–9; Braun, J., et al, *Environmental Health Perspectives*, 2009; 117 (12): 1945–52; Nanjappa, M. K., et al, *Biol Reprod*, 1 Feb 2012 (epub)

Chapter 10 'The gay gene'

1 Bonde, J. P., et al, *Epidemiology*, September 2011; 22 (5): 617–19

2 Li, D. K., et al, *Fertil Steril*, February 2011; 95 (2): 625–30

3 Ravnborg, T. L., et al, *Hum Reprod*, May 2011; 26 (5): 1000–11

4 Fowler, P. A., et al, *J Clin Endocrinol Metab*, 13 July 2011 (epub)

5 Prudic, K. L., et al, *Science*, 2011; 331 (6013): 73–5

6 Georges, A., et al, *Sex Dev*, 2010; 4: 7–15; Crain, D.A., et al,

Reproductive Toxicology, 2007; 24: 225–39

7 Wibbels, T., et al, *J of Experimental Zoology*, 1991; 260 (1): 130–4

8 Grech, V., et al, *J Epidemiol Community Health*, August 2003; 57 (8): 612–15

9 Bao, A. M., et al, *Frontiers in Neuroendocrinology*, 2011; 32: 214–26

10 Ogas, O., and Gaddam, S. A., *Billion wicked thoughts*, Dutton, New York, 2011

11 http://www.guardian.co.uk/world/2007/sep/26/iran.gender

12 Olyslager, F., et al, http://ai.eecs.umich.edu/people/conway/TS/Prevalence/Reports/Prevalence%20of%20Transsexualism.pdf

13 Segal, N., *Arch Sex Behav*, June 2006; 35 (3): 347–58

14 Schöning, S., et al, *J Sex Med*, May 2010; 7 (5): 1858–67

15 Manning, J. T., *Proc Natl Acad Sci USA*, 2011; 108 (39): 16143–4

16 Eisenberg, M. L., et al, *PLoS ONE*, 11 May 2011; 6 (5): e18973

17 Swan, S. H., et al, *Environ Health Perspect*, 2005; 113 (8): 1056–61

18 Manning, J. T., et al, *Human Reproduction*, 1998; 13: 3000–4

19 Voracek, M., et al, *Psychol Rep*, June 2009; 104 (3): 922–56; Voracek, M., et al, *Personality and Individual Differences*, 2011; 51, 417–22

20 Breedlove, S. M., *Endocrinology*, 2010; 151 (9): 4116–22

21 Bennett, M., et al, *J Sports Sci*, 2010; 28 (13): 1415–21; Hönekopp, J., and Schuster, M., *Personality and Individual Differences*, 2011; 48: 4–10

22 Ferdenzi, C., et al, *Proc Biol Sci*, 2011; 278 (1724): 3551–7

23 Choi, I. H., et al, *Asian J Androl*, 4 July 2011 (epub)

24 Paul, S. N., et al, *Twin Res Hum Genet*, 2006; 9: 215–19
25 Medland, S. E., et al, *Am J Hum Genet*, 2010; 86 (4): 519–25
26 Oinonen, K. A., *Psychoneuroendocrinology*, 2009, 34: 713–26
27 Sanders, G., et al, *Arch Sex Behav*, October 1997; 26 (5): 463–80; Hall, J. A., et al, *Archives of Sexual Behavior*, 1995; 24 (4): 395–407
28 Grov, C., et al, *Arch Sex Behaviour*, June 2010; 39 (3): 788–97
29 Jordan-Young, R. M., *Brain Storm: The Flaws in the Science of Sex Differences*, Harvard University Press, Cambridge, Mass., 2010
30 Sell, R. L., et al, *Arch Sex Behav*, June 1995; 24 (3): 235–48
31 Rosenthal, A. M., et al. *Biol Psychol*, September 2011; 88 (1): 112–15; Rosenthal, A. M., et al, *Arch Sex Behav*, 23 December 2011 (epub); Rieger, G., et al, *Psychological Science*, 2005; 16: 579–84
32 Långström, N., et al, *Arch Sex Behav*, February 2010; 39 (1): 75–80
33 Hamer, D., and Copeland, P., *The Science of Desire: The Search for the Gay Gene and the Biology of Behavior*, Simon & Schuster, New York, 1994; Hamer, D. H., et al, *Science*, 16 July 1993; 261 (5119): 321–7
34 Hu, S., et al, *Nat Genet*, 1995; 11 (3): 248–56; Rice, G., et al, *Science*, 1999; 23; 284 (5414): 665–7
35 LeVay, S., *Queer Science: The Use and Abuse of Research into Homosexuality*, MIT Press, Cambridge, 1996
36 Camperio-Ciani, A., et al, *Proc Biol Sci*, 7 November 2004; 271 (1554): 2217–21
37 Blanchard, R., *Horm Behav*, September 2001; 40 (2): 105–14

38 Green, R., *Psychological Medicine*, 2000; 30: 789–95
39 Blanchard, R., and Lippa, R., *Archives of Sexual Behavior*, December 2008; 37 (6): 970–6
40 Bogaert, A. F., et al, *Front Neuroendocrinol*, April 2011; 32 (2): 247–54
41 Struckman-Johnson, C., et al, *J Interpers Violence*, December 2006; 21 (12): 1591–1615
42 Stoller, R. J., et al, *Arch Gen Psychiatry*, April 1985; 42 (4): 399–404
43 Kendler, K. S., et al, *Am J Psychiatry*, 2000; 157 (11): 1843–6; Case, L. K., et al, *Med Hypotheses*, 2012, 78 (5), 626–31
44 http://www.alternet.org/story/146557/?page=entire
45 Diamond, L. M., *Sexual Fluidity*, Harvard University Press, 2008
46 Diamond, L. M., and Butterworth, M., *Sex Roles*, 2008; 59 (5–6): 365–76
47 Bailey, J. M., and Zucker, K. J., *Developmental Psychology*, 1995; 31: 43–55; Diamond, L. M., *Dev Psychol*, September 1998; 34 (5): 1085–95
48 Burri, A., et al, *PLoS ONE*, 2011; 6 (7): e21982
49 Långström, N., et al, *Arch Sex Behav*, 2010; 39 (1): 75–80
50 Bocklandt, S., et al, *Hum Genet*, 2006; 118 (6): 691–4
51 Diamond, M., et al, *Horm Behav*, 1996; 30 (4): 333–53
52 Crews, D., et al, *Proc Natl Acad Sci*, 2007; 104 (14): 5942–6
53 Farabollini, F., et al, *Environ Health Perspect*, 2002; 110 (Suppl3): 409–14; Quignot, N., et al, *Reprod Toxicol*, 21 Jan 2012 (epub)

Chapter 11 'The fidelity gene'

1 Prause, N., et al, *Archives of Sexual Behavior*, 2004; 36 (3):

341–56; Bogaert, A. F., *Journal of Sex Research*, 2004; 41: 279–87

2 Aron, E. N., and Aron, A. J., *Pers Soc Psychol*, August 1997; 73 (2): 345–68

3 Jagiellowicz, J., et al, *Soc Cogn Affect Neurosci*; Jan 2011; 6 (1): 38–47; Belsky J., and Pluess, M., *Psychol Bull*, 2009; 135 (6): 885–908

4 Killinsworth, M. A., and Gilbert, D. T., *Science*, 2011; 330: 932

5 Mark, K. P., et al, *Arch Sex Behav*, 11 June 2011

6 Cherkas, L. F., et al, *Twin Res*, 2004; 7 (6): 649–58

7 Brouwer, L., et al, *Mol Ecol*, August 2010; 19 (16): 3444–55

8 Forstmeier, W., et al, *Proc of the Nat Acad Sciences*, 2011; 108 (26): 10608–13

9 Mark, K. P., et al, *Arch Sex Behav*, 2011; 40 (5): 971–82

10 de Bono, M., et al, *Cell*, 1998; 94 (5): 679–89

11 Wang, Z. X., et al, *J Neuroendocrinol*, 2000; 12 (2): 111–20

12 Pitkow, L. J., et al, *J Neurosci*, 2001; 21 (18): 7392–6

13 Shahrokh, D. K., et al, *Endocrinology*, 2010; 151 (5): 2276–86

14 Walum, H., et al, *Proc Natl Acad Sci*, 2008; 105 (37): 14153–6

15 Hurlemann, R., et al, *J Neurosci*, 2010; 30 (14): 4999–5007

16 Waldinger, M. D., et al, *J Sex Med*, 2009; 6 (10): 2778–87

17 Winkelmann, J., et al, *Nature Genetics*, 2007; 39 (8): 1000–6

18 Dunn, K. M., et al, *Biol Lett*, 2005; 1 (3): 260–3

19 Dawood, K., et al, *Twin Res Hum Genet*, 2005; 8 (1): 27–33

20 Perry, J. D., and Whipple, B. J., *Sex Res*, 1981; 17: 22–39

21 Burri, A. V., et al, *J Sex Med*, 2010; 7 (5): 1842–52

22 Jannini, E. A., et al, *J Sex Med*, 2010; 7: 2292–4

23 http://www.newscientist.com/article/dn20770-sex-on-the-brain-what-turns-women-on-mapped-out.html

24 Graves, J. A., *Biol Reprod*, September 2000; 63 (3): 667–76
25 Lloyd, J., et al, *Br J Obstetrics Gynaecol*, 2005; 112: 643–6
26 Lloyd, E. A., *The Case of the Female Orgasm: Bias in the Science of Evolution*, Harvard University Press, Cambridge, Mass., 2005; Wallen, K., and Lloyd, E. A., *Evolution and Development*, 2008; 10: 1, 1–2
27 Komisaruk, B. R., et al, *J Sex Med*, 2011; 8 (10): 2822–30
28 Burri, A. V., and Spector, T., *J Sex Med*, 2011; 8 (9): 2420–30
29 Burri, A. V., et al, *J Sex Med*, 2009; 6 (7): 1930–7
30 Burri, A. V., et al, *J Sex Med*, 2012; 9 (1): 198–206
31 Hawkes, N., *BMJ*, 5 October 2010; 341: c5532
32 Goldfischer, E. R., et al, *J Sex Med*, 2011; 8 (11): 3160–70
33 Rowe, A. C., et al, *Psychoneuroendocrinology*, 2011; 36 (8): 1257–60
34 Meston, C. M., et al, *Annu Rev Sex Res*, 2004; 15: 173–257
35 Komisaruk, B. R., and Whipple B., *Annu Rev Sex Res*, 2005; 16: 62–86
36 Georgiadis, J. R., et al, *Hum Brain Mapp*, October 2009; 30 (10): 3089–3101
37 Komisaruk, B., Beyer, C., Whipple, B., *The Science of Orgasm*, JHU Press, 2006
38 Wilson, L. A., et al, *J Sex Med*, April 2009; 6 (4): 947–57; Komisaruk, B., and Whipple, B., *Annu Rev Sex Res*, 2005; 16: 62–86
39 Bell, J., et al , *PLoS Gen*, 2012; 8 (4): e1002629 (epub)

Chapter 12 Bacteria genes

1 Crew, K., and Neugut, A., *World J Gastroenterol*, 2006; 12 (3): 354–62
2 Lugli, A., et al, *Nature Clinical Practice Gastroenterology and*

Hepatology, 2007; 4: 52–7

3 Sonnenberg, A., et al, *Gastroenterology*, 2007; 132 (7): 2320

4 Dingle, K. E., *PLoS ONE*, 2011; 6 (5): e19993

5 Brown, L. M., *Epidemiol Rev*, 2000; 22 (2): 283–97

6 Strachan, D. P., *Thorax*, August 2000; 55 (Suppl 1): S2–S10

7 Blaser, M. J., et al, *Gut*, 2008; 57 (5): 561–7; Azad, M. B., and Kozyrskyj, A. L., *Clin Dev Immunol*, 2012; 932072 online

8 Wang, K. Y., et al, *Am J of Clinical Nutrition*, 2004; 80 (3): 737–41

9 Maslova, E., et al, Abstract Presented at European Respiratory Society Conference, 2011

10 Robinson, M. K., *N Engl J Med*, 2009; 361: 520

11 Vetter, M. L., *CML – Diabetes*, 2010; 27 (1): 1–12

12 Li, J. V., et al, *Gut*, September 2011; 60 (9): 1214–23

13 Furet, J. P., et al, *Diabetes*, December 2010; 59 (12): 3049–57

14 Casadesús, J., et al, *Microbiol Mol Biol Rev*, 2006; 70 (3): 830–56

15 Collier, J., *Curr Opin Microbiol*, December 2009; 12 (6): 722–9

16 Gonda, T. A., et al, *Gastroenterology*, 13 January 2012 (epub)

17 Sepulveda, A. R., et al, *Gastroenterology*, 2010; 138: 1836–44

18 Wang, Z., et al, *Nature*, 2011; 472 (7341): 57–63

19 Koren, O., et al, *Proc Natl Acad Sci USA*, 2011; 108 (Suppl 1): 4592–8

20 Bahekar, A. A., et al, *Am Heart J*, 2007; 154 (5): 830–7

21 D'Costa, V. M., et al, *Nature*, 2011; 477 (7365): 757–61

22 Mohammed, I., et al, *Am J Gastroenterol*, June 2005; 100 (6): 1340–4

23 Bengtson, M. B., et al, *Gut*, 2006; 55: 1754–1759

24 Relman, D. A., *N Engl J Med*, 2011; 365: 347–57

25 Arumugam, M., et al, *Nature*, 2011; 473 (7346): 174–80
26 Costello, E. K., et al, *Science*, 2009; 326 (5960): 1694–7
27 Holdeman, L. V., et al, *Appl Environ Microbiol*, 1976; 31 (3): 359–75
28 Qin, J., et al, *Nature*, 4 March 2010; 464 (7285): 59–65
29 Arumugam, M., et al, *Nature*, 12 May 2011; 473 (7346): 174–80
30 Turnbaugh, P. J., et al, *Nature*, 2006; 444: 1027–31
31 Bäckhed, F., *Proc Natl Acad Sci USA*, 2007; 104 (3): 979–84
32 McGarr, S. E., et al, *Curr Gastroenterol Rep*, 2009; 11 (4): 307–13
33 Moore, W. E., et al, *Appl Environ Microbiol*, September 1995; 61 (9): 3202–7; Kostic, A. D., et al, *Genome Res*, February 2012; 22 (2): 292–8; Castellarin, M., et al, *Nutr Rev*, September 2009; 67 (9): 509–26
34 van Baarlen, P., et al, *Proc Natl Acad Sci*, 2009; 106 (7): 2371–6; Kellermayer, R., et al, *FASEB*, May 2011; 25 (5): 1449–60
35 Marchesi, J. R., et al, *PLoS ONE*, 2011; 6 (5)
36 Neufeld, K. M., et al, *Neurogastroenterol Motil*, 2011; 23: 255e264, e119; Heijtz, R. D., et al, *Proc Natl Acad Sci*, 2011; 108: 3047e3052; Lyte, M., et al, *Physiol Behav*, 2006; 89: 350e357
37 Bravo, J. A., et al, *Proc Natl Acad Sci*, 29 August 2011 (epub)
38 Karczewski, J., et al, *Am J Physiol Gastrointest Liver Physiol*, 2010; 298 (6): G851–6; Kau, A. L., et al, *Nature*, 2011; 474 (7351): 327–36
39 Grehan, M. J., et al, *Journal of Clinical Gastroenterology*, 2010; 44 (8): 551–61
40 Borody, T., *Am J Gastroenterol*, 2000; 95 (11): 3028–9
41 Pimentel, M., et al, *N Engl J Med*, 6 Jan 2011; 364 (1): 22–32

42 Guilloteau, P., et al, *Nutr Res Rev*, 2010; 23 (2): 366–84

43 Blaser, M., *Nature*, 2011; 476: 393–4; Cho, I., and Blaser, M. J., *Nat Rev Genet*, 2012; 13 (4): 260–70; Clemente, J. C., et al, *Cell*, 2012; 148: 1258–70

44 Pauline, T., et al, *Biomed Environ Sci*, 2001; 14 (3): 207–13

45 Jia, J., et al, *FEMS Microbiol Ecol*, 2011; 78 (2): 395–404

46 Roger, L. C., et al, *Microbiology*, 2010; 156 (Pt 11): 3317–28

Chapter 13 Identical genes

1 Gurdon, J. B., et al, *Nature*, 1958; 182 (4627): 64–5

2 Takahashi, K., and Yamanaka, S., *Cell*, 2006; 126 (4): 663–6

3 Ebben, J. D., et al, *World Neurosurg*, 2011; 76 (3–4): 270–5; Makkar, R. R., et al, *Lancet*, 14 February 2012 (epub)

4 Berdasco, M., and Esteller, M., *Stem Cell Res Ther*, 31 October 2011; 2 (5): 42

5 Reefhuis, J., et al, *Hum Reprod*, 2009; 24 (2): 360–6

6 Strawn, E. Y., et al, *Fertil Steril*, July 2010; 94 (2): 754. e1–2

7 Rajender, S., et al, *Mutat Res*, May–Jun 2011; 727 (3): 62–71

8 McDonald, S., et al, *Am J Obstet Gynecol*, 2005; 193: 141–52; Ombelet, W., et al, *Hum Reprod*, 2006; 21: 1025–32

9 Li, L., et al, *Fertil Steril*, May 2011; 95 (6): 1975–9

10 Gordon, L., et al, *Epigenetics*, May 2011; 6 (5): 579–92

11 Edwards, R. G., and Beard, H. K., *Hum Reprod Update*, Nov–Dec 1998; 4 (6): 791–811

12 Hashiloni-Dolev, Y., et al, *Public Understanding of Science*, 19 (4): 435–51

13 Prainsack, B., et al, *Hum Reprod*, 2007; 22 (8): 2302–8

14 Prainsack, B., and Spector, T. D., *Soc Sci Med*, 2006; 63 (10): 2739–52

15 Wilmut, I., The General Assembly of the United Nations, in its 'Declaration on Human Cloning' (United Nations' General

Assembly, 2005)

16 Prainsack, B., and Spector, T. D., *Soc Sci Med*, 2006; 63 (10): 2739–52

17 Huck, S., et al, *The Economic Journal*, 2005; 505: 689

18 Edelman, G. M., *Neuron*, February 1993; 10 (2): 115–25; McDowell, J. J., *Behav Processes*, May 2010; 84 (1): 358–65

19 Baggini, J., *The Ego Trick*, Granta Books, 2011

20 Prainsack, B., and Kitzberger, M., *Soc Stud Sci*, February 2009; 39 (1): 51–79

21 Fox Keller, E., *The Mirage of a Space between Nature and Nurture*, Duke University Press, 2010

INDEX

Abbot, Jacob, 113
absolute pitch (AP), 80–1
acid secretion (H2) blockers, 272
Actiheart study, 78
Acylin Therapeutics, 170
advertisements, 52
Afghanistan, 138, 234
Alabama research group, 86
Ali, Muhammad (formerly Cassius Clay), 82
Alien, 177
Amazon, 47, 182–3
Amish birth rate, 103–4
Andretti, Mario, 83
Angelman syndrome, 214
anorexia nervosa, 66–7
antibiotics, 277, 268, 276
antidepressants, SSRI, 212
antioxidant supplements, 190–1
Arabidopsis, 35–6
Aryan Kultur settlements, 77
Asperger's syndrome, 204, 205, 209–10
athletes, 68, 72–3
Atkins, Dr, and diet, 183
Atlas Sports Genetics, 67
Australian twins, 51
autism and Autism Spectrum Disorder (ASD), 40, 204–217
Averill, Meredith, 186–7

bacteria, 260, 262–3
  Bifidobacillus, 265, 274
  Helicobacter pylori (HP), 272–3, 264, 265, 267
  Lactobacillus, 265, 274
'bad genes', 130–52
Barbie doll adverts, 221
Baron-Cohen, Simon, 145, 210, 212–13
Barr, Rosanne, 170
Battle Hymn of the Tiger Mother, 70, 112
BBC, 112, 181
Beatles, The, 71
Beaver, Kevin, 151–2
Beckwith-Wiedemann syndrome, 282
bees, 221
Beijing Genomics Institute, 272–3
Bell, Mary, 171
Belsky, Jay, 151–2
beta-carotene, 190, 191
Bettelheim, Bruno, 115
BGI, 282–3
Bifidobacillus bacteria, 265, 274
Binet, Alfred, 83
BioArts International Inc, 280
Biology, 31
birds, 244–5
Birmingham, 134
bisphenol A (BPA), 216–18, 220, 221, 238
Blackadder, 51
Blaine, David, 181
BMI (body-mass index), 167

Boehringer, 255–6
Bonaparte, Princess Marie (Dr A. E. Narjani), 252–3
Bonaparte, Napoleon, 261
Borody, Professor Tom, 275–6
Boston, Playland club, 223
Bouchard, Tom, 43
bowel cancer, 284
bowel syndrome, irritable (IBS), 268–9, 270, 274–5, 276
Boys from Brazil, The, 288
BPA (bisphenol A), 216–18, 220, 221, 238
brain, changes in, 43, 98
brain areas
  after strokes or injuries, 208
  amygdala, 213–14
  anterior cingulate gyrus, 258
  dentate gyrus, 85
  hippocampus, 74, 147, 208
  hypothalamus, 168, 169
  inferior parietal lobe, 208
  left orbito-frontal complex (OFC), 258
  prefrontal cortex (PFC), 258
brain hormone, glucocorticoid receptor, 126–7, 148
brain scans, MRI, 147, 149
brain waves, mu, 218
brains, abnormal metabolism in, 125
Brazil, epilepsy suffers in, 98
breast cancer, 192–3, 194–5, 196, 197, 198–9, 201
  'alternative' medications, 196–7
  and fat intake, 196
  increase in, 195
  response to treatment, 199
Briand, Aristide, 253
broken mirror syndrome, 206, 207
Brussels, St Luc hospital, 166
Bucharest orphanages, 122
Buddhist monks, 64
Budleigh Salterton, 130–1, 132
  St Peter's church, 130
bullying in schools, 128
Burge, David, 80
Burri, Andrea, 252, 253
Burt, Sir Cyril, 41, 42, 43
butterflies, 37, 221
butyrate, 276
Byron, Ruth and Simon, 247
Byzantine Empire, 132

calcium, 191
California, autism in, 211, 215
Californian gifted children IQ study, 57, 83
Cambridge, 212, 213
Campbell, Keith, 279
Canada, BPA ban in, 218
Canadian calorie restriction diet, 185
cancer, 192–202, 265 *see also individual entries*
Caplan, Bryan, 112, 115
Carlson, Dr Mary, 123
Case of the Midwife Toad, The, 31
cataracts, 39–40
caterpillars, 37
Catholic Church, 137
cats, 290–1
Ceausescu, Nicolae, 122–3
Cézanne, Paul, 73–4
characteristics, acquired, inheritance of ('soft inheritance'), *see* epigenetics
cheese protein, 174
Chegwin, Keith, 83
Cherkas, Lynn, 253
chickens, 37–8
child abuse, 130–3, 134–9, 146, 148–9
child prodigies, 69–71, 76
childbirth, 'orgasmic', 258–9
children
  badly abused and neglected, MRI scans of, 149
  bullying of, 128

differing views on upbringing, 112, 113–15, 128–9
from failing families, 150–1, 152
religious encouragement to have, 103–4
rise of behavioural and psychiatric disorders in, 111
in 17th or 18th century, 110–11
underlying genetic backgrounds, 125–6
China, sperm count study in, 220
choline, 176, 177, 267
chromosomes, 197
chromosome 15: 214
X, 214, 232–3, 238, 290
Y, 144, 146, 221–2, 251
chronic fatigue syndrome, 269–70
Chua, Amy, 70, 112, 113
circumcision, male, 101
Clark, Dick, 203, 204
Clay, Cassius (later Muhammad Ali), 82
Cleopatra, 101
Clinton, Bill, 96
clitoris, 251, 252, 253
Clone, 290
cloning, 278–9, 280–1, 281–2, 283–4, 288
Clostridium difficile, 263–4, 275
CODIS system, 143
coeliac disease, 274
cognitive behavioural therapy (CBT), 64, 138, 210
Cold Case, 142
Collins, Francis, 96
Collins, John, 143–4
colon cancer, 200, 274
Colvin, Geoff, 70
conception, 214, 215
Connery, Sean, 242
contraceptive pill, 195–6
Cooper, Carrie, 141
Cooper, George and Henry, 82–3
Cooper, Jerome and Tyrone, 140–1
CoppaFeel charity, 194
Cornell Univeristy, 272
Coyle, Daniel, 70
cress, thale, 35–6
criminality, 145–6
Crohn's disease, 274, 275
CSI, 142
cultural trends and personality traits, 99–101
Cuvier, George, 27–8

Dacogen, 200
Daily Star, 49, 52
Daily Telegraph, 40
dairy (lactose) intolerance, 172, 173, 174
Dance, Colin and Maurice, 131
Dance, Tommy, 130–1
Dance, William, 130, 131, 133
Danish pregnancy study with low-fat yogurts, 265
Daphnia, 36
Darwin, Charles, 9, 27, 28, 29, 77, 88, 96, 105, 221, 261, 293
Darwinism, Neural, 289
Dawkins, Richard, 98, 109, 291–2
Day, Binny, 130, 131
Day, Carl, 131, 133
Deer, Brian, 212
Denmark, 53, 54
Dennett, David, 97
dental caries (tooth decay), 267–8
Denver, 150
Depp, Johnny, 242
depression, heritability of, 59–62
DES (diethylstilboestrol), 201, 202
diabetes, 190, 265
diet, 181–9
diet alterations and fur colour, 38, 176–7, 216
diethylstilboestrol (DES), 201, 202
digit ratios, 2D:4D, 226–8, 229, 242, 237
distorted tunes test, 79

Djerassi, Carl, 195
DNA, 9, 17, 34, 140–3, 278, 280, 290–1
dogs, 280, 281
Dolly the sheep, 288–9, 280, 283
dopamine, 127, 147–8
Drayna, Dennis, 79
drugs in sport, 87
Dublin priests, 102
Duchovny, David, 241
Duke University, 38, 176–7
Dundee, Angelo, 82
Dunedin Study, 126, 146
Dutch famine, 1944, 188
Dweck, Carol, 62–3

E. coli outbreak, German salad, 263
Eastenders, 131
Eaves, Lindon, 92–3
Ecstasy (MDMA), 246
Edwards, Jonathan, 81
Edwards, Robert, 282
Einstein, Albert, 87
Eldoret, Kenya, 72
Elmira, New York, 150
empathy, 145, 147–8
England, 13th and 14th century, diet in, 181–2
English and Romanian Adoption Study (ERA), 124–5, 126
Enlightenment, 77
enterotypes, 273
environmentalists, 40–4, 62
enzyme, monoamine oxidase-A (MAOA), 146
epigenetic changes related to environment, 238–9
epigenetic drugs, 87, 149, 199–200
epigenetic influence in ASD, 214
epigenetic signals, 152
epigenetics ('soft inheritance'), 27, 28, 29, 30–1, 32, 33, 34–9, 293
  in the future, 290–3
  histone acetylation, 170
  as key to cancer-forming process, 197–8, 199–200, 201
  and muscle cells/growth, stimulating, 85–6
  nature vs. nurture, 39–44
epilepsy, 98
Ericsson, Anders, 70, 71, 75–6, 78
Ethiopians, 73
eugenics, 77
Eugenics Society, 41
Eurovision song contest, 69
'evil genes', 139–43
examinations, 11-plus, 41
experience and talent, 73–8
Eysenck, Hans, 49–50

faith, 91
faith, losing, 106–9
'faith gene', 91–5, 108
families, failing, 149–52
Far Side cartoons, 48–9, 50
'fat genes', 166–92
fatigue syndrome, chronic, 269–70
fecal transplant trial, 274–5
'feederism', 166
fibromyalgia, 136, 269
fidelity, 240–59
fidgeting, 184
finger length, 226–8, 229, 242, 237
Finnish families study, 80
fish oils, 191
flame retardants, BDE-4, 217
fleas, water, 37
flibanserin, 255–6
Flint, Larry, 100
Florida, obesity in, 167
Flowering Locus C gene, 36
folate, 176
folic acid supplements, 177, 190, 266–7
food allergies, 172–3
food deprivation, 188
Ford, Gina, 114, 115
Fowler, James, 64

fragile X syndrome, 214
Frederick II, Emperor, 111, 124
French, Dawn, 170
Freud, Sigmund, 29, 253
Frimley Park Hospital, 270–1
Fritzl, Josef, 145
future developments, 289–3

G spot, 251
GABA neurotransmitter receptors, 275
Gallup world happiness survey 2011, 53–4
Galton, Francis, 77–8
gastric surgery, 265–6
Gattaca, 289, 290
Gebrselassie, Haile, 73
Geek2Geek, 213
'geeks', 212, 213
gene sequencing machine, 263–4
General Medical Council (GMC), 212
genes
  ACE (Acetyl CholinEsterase – 'sports endurance'), 67, 68
  ACE II, 67
  ACTN3 (alpha actinin-3 – 'running'), 68–9, 72–3
  Agouti, 38, 176, 216
  amylase, 169
  bacterial, 266–7, 268, 271–2
  BRCA1 mutations, 194–5, 197
  BRCA2 mutations, 194–5
  cell hosting, 292
  CNV, 205
  copies (alleles), 38–9
  dopamine, 126
  E-cadherin, 200
  5-HTT variant, 60–1
  Flowering Locus C, 37
  FTO, 168–9, 171
  HER2, 199
  identical, 278–94 *see also* 'cloned/cloning' *entries*; twins, identical
  imprinted, 38–9, 171
  insect, modification of, 36
  ISR1, 170
  KLF14 obesity master regulator, 170–1
  Lcyc, 35
  neuropeptide Y, 245
  passenger, 263
  protocadherins, 127
  PTC, 174
  retinoid receptor, 185
  SEPT6, 200
  serotonin, 144
  sirtuin family, 187–8, 188
  SRY gender-determining, 221–2
  'talent', *see* 'talent genes'
  vasopressin receptor, 245–6
  VMAT2, 104
Genetic Saving, 280
German salad *E. coli* outbreak, 263
German study on shaving pubic hair, 99
'giggle gene', 46–52
Gilbert, Dan, 243–4
Gillie, Oliver, 41
giraffes, 28
girls, childhood development of, 221
Gladwell, Malcolm, 70, 71
GMC (General Medical Council), 212
GOAT (Ghrelin O-Acetyl Transferase) inhibitor, 170
Goldacre, Ben, 190
golfers, talented, 71–2
Graduate, The, 216
Gräfenberg, Dr Ernst, 251
Grand Rapids, Michigan, 139–41
Greece, 167
Gretzky, Wayne, 81
Guarente, Leonard, 187
Guildford, Barley Mow restaurant near, 270
Guinness Book of World Records, 69, 168

gum inflammation (periodontitis), 267–8
Gurdon, John, 279–80
gut bacteria, 264, 265, 266, 267, 272, 273–4, 276
gut flora, 265, 267, 272–3, 274, 275, 276
Gypsies, 123

haemorrhoids, 39–40
Haiti, 72
Hall, Jerry, 83
Hamer, Dean, 232–3, 238
happiness, 40, 45–65
happiness surveys, 53–4, 243–4
Harlow monkeys experiment, 123, 126
Harris, Judith Rich, 115
Harvard, 243–4
Hawke, Ethan, 290
Heiman, Julia, 257
Herceptin, 198, 199
Heritage study, 79
Hitler, Adolf, 288
homosexuality, 221–39
hormone disruptors, 202, 216–19, 238
hormones
  as cause of increase in ASD, 213–14, 216
  'cuddle', *see* oxytocin
  food regulating, 169, 170
  Ghrelin, 169, 170
humour, 47–52
Hungary, church attendance in, 103
hunger, hormones controlling feeling of, 169, 170, 184
Huntington's disease, 292
Hutterites, 104
Hwang, Dr Woo-suk, 281
'hygiene' hypothesis, 264–5

IBS (irritable bowel syndrome), 268–9, 270, 274–5, 276
Iliescu, Ion, 123
inheritance of acquired characteristics ('soft inheritance'), *see* epigenetics
insects, 36, 37
insulin, 178
International Athletics Association, 225
iPhone app, 243–4
IQ, 40, 41–2, 43, 63, 78, 83–4
IQcuties, 213
Iran, 212
Iraq veterans, 138
Ireland, decline of church attendance in, 101–2
Ireland, depression in, 59, 60
Irish priests, 102
irritable bowel syndrome (IBS), 268–9, 270, 274–5, 276
Island, The, 283
isoflavones, 196–7
Istodax, 200
IVF (in vitro fertilisation) treatment, 282–3
  from single genome, 278–9

Jackson, Georgia, 144
James, Oliver, 114–15
Jedi 'religion', 88
Jesuit order, 76
Jews, Orthodox, 103–4
Joan of Arc, 91, 98
John Hopkins University, 197–8
John Paul II, Pope, 261
jokes, 50
Joyce, James, 261

Kalenjin tribe, 72
Kamin, Leon, 41
Kammerer, Paul, 29–31, 37
Keller, Helen, 109
Kenyan athletes, 68, 72–3, 84–5
Khomeini, Ayatollah, 261
Killinsworth, Matthew, 243–4

Kinsey Institute, Indiana, 257
Knowledge, The, 74
Koestler, Arthur, 31
kombucha ('health' tea), 276

lactase, 173–4
Lactobacillus bacteria, 265, 274, 275
lactose intolerance, 172, 173, 174
Lamarck, Jean-Baptiste-Pierre-Antoine de Monet, Chevalier de, 27–9, 30, 33–4, 293
Lancet, The, 211–12, 262–3
Larson, Gary, 48–9, 50
Laszlo (Hungarian educational psychologist) and family, 76
Laura, Dr, 137
Leach, Penelope, 114–15
Lebensborn project, 77
legs, restless, syndrome (RLS), 247, 248
LeVay, Simon, 221
Lewis, C. S., 96
lignans, 196
Little Britain, 51
Locke, John, 76–7, 113
London
  Bellingham amateur boxing club, 82
  Hammersmith Hospital, 45
  Institute of Neurology, 74
  Institute of Psychiatry, 61, 124, 147
  King's Cross station, 45–6
  Lambeth, 106
  St Thomas' Hospital, 45
  taxi drivers, 74, 75, 208
  Wembley Stadium, 82
longevity, 153–165, 187–9
Louisiana families study, 79
Lucas, George, 88
Lykken, David, 55
Lysenko, Trofim, 31–2, 33, 36

Mad Men, 52, 53
Mahler, Alma, 30
maize hybrids, breeding of, 33
Malaysia, 'sissy camps' in, 235
Manning, John, 227
Marshall, Barry, 262–3, 276
marshmallow experiment, 180
Martin, Nick, 92–3
McDonald's diet, 178
McGlothin, Paul, 185–6
McGuire, Billy and Benny, 168
McIlroy, Rory, 71–2
MDMA (Ecstasy), 246
Meaney, Mike, 126
measles, 212
Medical Research Council (MRC), 215, 218
meditation, 64
'memes', concept of, 98–9
memory, 75–6, 86, 289
Memphis, 150
Mendel, Gregor, 9, 42
metagenome, 263, 272–3
methyl donors, 149
methylation, 34–6, 85, 86, 148, 151, 177, 185, 186, 266
  and cancer, 197–8, 199, 200, 214, 215, 216–17
mice
  Agouti, 38, 176, 216–17
  antibiotic study on, 267
  dietary experiments on, 178, 275
  fecal study of, 273
  folic acid study with, 266–7
  longevity research on, 187–8
  male deer, BPA study on, 217
  mutant, 68
microbes, 263, 271–2
Mincu, Dr Iulian, 122
mindsets, genetic, 62–5
mineral supplements, 190
Ministry of Agriculture, 215
Minnesota adopted-twin study, 22, 43, 45, 46, 55, 92, 116
mirror neurones, 206, 207, 258

mirror therapy, 207
Mischel, Walter, 180
mitochondrial activity, 188
MMR (measles, mumps, rubella) vaccine, 211–12
Mobley, Tony, 143–4, 145
 family, 144, 145
monkeys, 61, 206, 217
Monroe, Marilyn, 170
Montreal research team, 126–7, 148
moodscope.com, 64
MORI 2007 poll, UK, 52
Mormons, 104
mouth bacteria, 267–8
Mozart, Leopold, 70
Mozart, Wolfgang Amadeus, 70–1, 80
MRC (Medical Research Council), 215, 218
MRI brain scans, 147, 149, 258
MRSA (Methicillin-resistant *Staphylococcus aureus*), 263–4
Muhammad Ali (formerly Cassius Clay), 82
Multiplicity, 283
music, 79, 80
MuTHER Consortium, 170

Napoleon Bonaparte, 261
Narjani, Dr A. E. (Princess Marie Bonaparte), 252–3
NASA scientists, 272
National Health and Nutrition Examination (NHANE) study (US), 218
National Institutes of Health (NIH) (US), 232–3, 238
Nature, 30, 67
NEAD (non-shared environment in adolescent development) study, 119–21
'nerd' mating theory, 212, 213
Nerd Passions, 213
Netherlands children with 'pubophobia', 99–100
neurones, mirror, 206, 207, 258
New Englanders, 64–5
New Guinea, Sambia tribe in, 234
New Testament, 91
New York
 cab drivers, 74
 Carnegie Hall, 112
 Sloane Kettering Cancer Center, 199
New York Times, 30
New Zealand, male criminality study in, 146
New Zealand forensic service, 142, 143
Newton, Isaac, 42
NHANE (National Health and Nutrition Examination) study (US), 218
NIH (National Institutes of Health) (US), 232–3, 238
nipples, 251
Nobel, Alfred, 261
Noble, G. K., 30
Norwegian study of twins with IBS, 270
Nueva Germania, Paraguay, 77
Nurse Family Partnership (NFP), 150
Nurture Assumption, The, 115
nurture 'blank slate' concept, 76–7

Obama, President Barack, 150
obesity, 12,
 extreme, 166
 and gastric bypass/banding surgery, 265–6
 growth of, 166–7, 184, 220
 and gut bacteria, 273–4
 heritability of, 167–8
obesity clinics, 166
oesophagus, cancer of, 265
oestrogen, 196, 201, 226–8
Office, The, 51–2

Olds, David, 150
olms (cave-dwelling salamanders), 29
$100,000 Pyramid, 203
optimism, 56–7, 58–9, 61
orgasms, female, 247–8, 249–54, 255–6, 257, 258–9
osteoarthritis, premature, 279
Owen, Clive, 242
Oxford University, 135, 194
oxytocin ('cuddle hormone'), 127, 147–8, 168–9, 245–6, 259
oxytocin nasal spray, 246, 256

pain relief and orgasms, 259
Palmer, William, 223–4
paracetamol (acetaminophen), 215
paranormal, beliefs in the US in, 108
parenting, books on, 70, 111, 112, 113, 114, 115
parenting, 110–29, 133–4, 149–51, 152
parents and guilt, 112–18
Pashtun men, 234
Peck, Gregory, 288
Peltonen, Leena, 78
penis size, 227–9, 253
peptic (stomach) ulcers, 261–3
perfectpitch.com, 80–1
periodontitis (gum inflammation), 267–8
Perry, Bruce, 149
personal identity, 289
personality disorders, 40
personality feature studies, 56
personality traits and cultural trends, 99–101
pessimism, 57, 59, 61, 64
pets, cloned, 280–1
phyto-oestrogens, 196, 216
Picasso, Pablo, 73–4
piglets, cloned, 291
pitch, absolute/perfect (AP), 80–1
plants, epigenetic changes in, 34–6
plastics, 216–19 *see also* BPA
Playboy, 100
Plomin, Robert, 119, 121
Positive Psychotherapy Intervention, 63–4
post-traumatic stress disorder (PTSD), 136, 138–9
Prader-Willi syndrome, 214
Prainsack, Barbara, 284
proteins, TET, 200
Prozac, 211
PTSD (post-traumatic stress disorder), 136, 138–9
pubic hair, 99–101

Quran, 91, 94

Ramachandran, V. S., 206, 207
rats
  breast cancer studies on, 201
  hormone disruptors reducing sexual attraction in, 238
  male, epigenetic effects on offspring's health, 178
  maternal licking and grooming experiments, 126–7, 148
  memory research on, 86
  pregnant mothers' epigenetic effects on daughter's and granddaughters' genes, 177
  and running, 85
Reiss, David, 120
religion, 88–109, 235
restless legs syndrome (RLS), 247, 248
resveratrol, 189, 202
Rett syndrome, 214
rifaximin, 276
Rift Valley, 68, 72
Rind, Dr, 137–8
Rizzolatti, Giacomo, 206
RNA, non-coding, 248
RNL Bio, 291
Rock, John, 195–6

rodents, experiments, 126–7, 148
Romania, 122, 123–5, 126
Rossellini, Isabella, 83
Rousseau, Jean-Jacques, 113
Royal Institution Christmas Lecture, 175
running, 68–9, 72–3
Russia, religion in, 102–3, 108
Rutter, Michael, 124

SAHA histone-modifying drug, 292
St Ignatius Loyola, 76
St Louis fecal study, 273
salamanders, cave-dwelling (olms), 29
Sambia tribe, 234
schizophrenia, 40
Science, 30
Sears, Dr Bill and Martha, 114–15
Seligman, Marty, 63–4
Semenya, Caster, 225
sensory processing sensitivity (SPS), 243
serotonin, 60–1, 147–8
Seventh Day Adventists, 175–6
sex addiction (hypersexual disorder), 240–1
sex-change operations, 223, 224
sex determination and development, 221–2, 225, 227
sexual abuse of children, 130–3, 134, 146
  common effects of, 136–8, 139, 146, 148–9
  victims, 134–9
sheep, Dolly the, 278–9, 280
Shenk, David, 70
Silicon Valley, 212
Simpson, O. J., 141, 240
skin colour, 73
Skinner, Mike, 238
Sleeper, 283
Sloane Kettering Cancer Center, 199
smoking, maternal, 220
Southampton diet study, 185
Soviet Union, 32, 33, 102–3, 108
soya beans, 202
soya isoflavones, 196–7
Space: 1999, 210
Spock, Dr Benjamin, 111, 114
sport, drugs in, 87
sporting abilities, 78
'sports endurance gene', 67, 68
Stalin, Josef, 32
Stepford Wives, The, 283
Stockdale, Vice Admiral James Bond, 58
Stockdale paradox, 58–9
stomach (peptic) ulcers, 261–3
stomach cancer, 260–1, 265, 266–7
Strachan, David, 264–5
Strange, Curtis, 83
stresses, life, 61
suicide, 59–60, 147
Sulston, Sir John, 175
Summer, Daniel, 145
Sun, 254
Sunday Times, 42, 200
supplements, mineral and vitamin, 190
Surrogates, The, 283
Sweden, 146, 191
Syed, Matthew, 70, 81

tadpoles, 279–80
talent, 66–87, 208
Tamoxifen, 193
Tampa, 204
Tanzania, 55
taxi drivers, London, 74, 75, 208
temperature changes, effect of, on species development, 221
Terman, Lewis, 83
Terman Study, 57
testicular cancer, 220
testosterone, 213, 220, 222, 225, 229, 233, 251–2

fetal, assessing, 226–8, 229
Thomas Aquinas, 96
toadflax, common, 34–5
toads, midwife, 29, 30–1, 36
tone-deafness, 79–80
tooth decay, 267–8
transsexuals, 223–4
Treffert, Dr Darold, 204
triclosan, 218
Tromovitch, Dr, 137–8
Trudghill, Nigel, 270
Turkheimer, Eric, 119
Twain, Mark, 73–4
Twin Registry, UK, 42–3
twins
absolute pitch study, 80
adopted, and religious upbringing, 88–90, 92
adopted, brought up in different surroundings, 115–18, 128
with Asperger's syndrome, 209–11
Australian, 51
autistic savant, 202–4
autistic studies of, 205, 214, 215
average heritability of, 78
and belief in God studies, 92
and bullying, 128, 133–4
and cancer, 192–5, 198–9, 200
coincidences between, 46–7
Cooper boxing, 82–3
and depression, 59–60
differing Ghrelin hormone levels in, 169
with differing religious beliefs, 94–5, 106–7
digit ratio studies on, 228, 230, 237
dishonourable discharge from US military of, 145–6
DNA matches and, 140–1, 143
female, and orgasms, 249–51, 255, 257
and fidelity, 245
gut flora study in, 272–3
and happiness study, 53, 54–5
heritabilities for aggressive behaviour and psychopathic personality, 146, 147
with irritable bowel syndrome (IBS), 268–9, 270
motivation study of, 84
MZ, 285–6
in obesity research, 170, 171–2, 174, 179–80, 184, 185
one being bisexual, 236–7
one being gay, 229–32, 235
produced by IVF, 283
reuniting of, 45–7
similarity in weight of, 167–8
and sporting abilities, 78
studies of, 39–43, 45, 46, 47 *see also* humour
talented, with non-famous twin siblings, 83
in taste experiment, 175
tone-deafness test on, 79–80
transsexual, 222–4, 225
as victims of abuse, 134–6, 137, 138, 139
in willpower experiment, 181
world's heaviest, 168
twins, identical, 284
and bowel habit study, 270–1
female, with differing attitudes to men, 241–43
and pain relief studies, 259
in same classroom study, 121–2
views on twins of, 284, 285–8
Twinsburg, Ohio, annual festival, 51

United Kingdom
breast cancer in, 195
census 2001, 88
cost of children with problems and in care, 118
decline of religion in, 102
diet in 13th and 14th century England, 181–2

United Kingdom—*contd*
Muslim population in, 104
obesity in, 166–7
religious people as voters in, 108
United Nations, 284–5
Convention on the Rights of the Child, Article 19: 118
United States
beliefs in the paranormal in, 108
breast cancer in, 195
breeding of maize hybrids in, 33
Food and Drug Administration (FDA), 256
Food Standards Agency, 218
General Social Survey, 245
military, dishonourable discharge of twins from, 145–6
NEAD (non-shared environment in adolescent development) study, 119–21
obesity in, 166–7, 184
religious beliefs and practices in, 103
United States Army CSF (comprehensive solder fitness) programme, 64
United States Congress, 137–8
US Masters, 71, 72
US Open golf tournament, 2011, 71–2

Vanessa Mae, 69–70
Vargas, Alexander, 31
vegetarian diet, 175–6
vernalisation, 32, 33, 36
Viagra, 255
Vidaza, 212
Vienna, University of, 31
Vietnam War, 58
vinclozolin, 217, 238
vitamin A, 190
vitamin B12, 176
vitamin E, 190
vitamin levels before birth, 215
vitamin supplements, 190–1
vitamins at time of conception, 215
voles, 245, 246

Waddington, Conrad, 33
Wakefield, Andrew, 212
Wang Jun, 273
Ward, Paul, 75–6
Ward, William Arthur, 59
Warren, Robin, 262–3
Wessely, Professor Simon, 139
willpower, 180–1
Wilmut, Professor Ian, 279, 283, 284
Wilson, E. O., 97–8
wine consumption, 188–9, 191
Wonder, Stevie, 80
Woods, Tiger, 71, 240
Woo-suk Hwang, Dr, 291
World Values Survey 1984 to 2004, 103
worms, 187, 245
Wright, Sewell, 176

X-Factor, The, 74

Yamanaka, Shinya, 291
'Yankee Doodle Dandy', 79
yeast, male and female forms in, 238
yogurts, 'gut-friendly', 265, 275, 276

Zolinza, 200
Zoloft, 211